Principles, Process and Practice of Professional Number Juggling

Principles, Process and Practice of Professional Number Juggling (Volume 1 of the Working Guides to Estimating & Forecasting series) sets the scene of TRACEability and good estimate practice that is followed in the other volumes in this series of five working guides. It clarifies the difference between an Estimating Process, Procedure, Approach, Method and Technique. It expands on these definitions of Approach (Top-down, Bottom-up and 'Ethereal') and Method (Analogy, Parametric and 'Trusted Source') and discusses how these form the basis of all other means of establishing an estimate.

This volume also underlines the importance of 'data normalisation' in any estimating procedure, and demonstrates that the Estimating by Analogy Method, in essence, is a simple extension of Data Normalisation. The author looks at simple measures of assessing the maturity or health of an estimate, and offers a means of assessing a spreadsheet for any inherent risks or errors that may be introduced by failing to follow good practice in spreadsheet design and build.

This book provides a taster of the more numerical techniques covered in the remainder of the series by considering how an estimator can potentially exploit Benford's Law (traditionally used in Fraud Detection) to identify systematic bias from third party contributors. It will be a valuable resource for estimators, engineers, accountants, project risk specialists as well as students of cost engineering.

Alan R. Jones is Principal Consultant at Estimata Limited, an estimating consultancy service. He is a Certified Cost Estimator/Analyst (US) and Certified Cost Engineer (CCE) (UK). Prior to setting up his own business, he has enjoyed a 40-year career in the UK aerospace and defence industry as an estimator, culminating in the role of Chief Estimator at BAE Systems. Alan is a Fellow of the Association of Cost Engineers and a Member of the International Cost Estimating and Analysis Association. Historically (some four decades ago), Alan was a graduate in Mathematics from Imperial College of Science and Technology in London, and was an MBA Prize-winner at the Henley Management College (… that was slightly more recent, being only two decades ago). Oh, how time flies when you are enjoying yourself.

Working Guides to Estimating & Forecasting

Alan R. Jones

As engineering and construction projects get bigger, more ambitious and increasingly complex, the ability of organisations to work with realistic estimates of cost, risk or schedule has become fundamental. Working with estimates requires technical and mathematical skills from the estimator but it also requires an understanding of the processes, the constraints and the context by those making investment and planning decisions. You can only forecast the future with confidence if you understand the limitations of your forecast.

The Working Guides to Estimating & Forecasting introduce, explain and illustrate the variety and breadth of numerical techniques and models that are commonly used to build estimates. Alan Jones defines the formulae that underpin many of the techniques; offers justification and explanations for those whose job it is to interpret the estimates; advice on pitfalls and shortcomings; and worked examples. These are often tabular in form to allow you to reproduce the examples in Microsoft Excel. Graphical or pictorial figures are also frequently used to draw attention to particular points as the author advocates that you should always draw a picture before and after analysing data.

The five volumes in the Series provide expert applied advice for estimators, engineers, accountants, project risk specialists as well as students of cost engineering, based on the author's thirty-something years' experience as an estimator, project planner and controller.

Volume I Principles, Process and Practice of Professional Number Juggling
Alan R. Jones

Volume II Probability, Statistics and Other Frightening Stuff
Alan R. Jones

Volume III Best Fit Lines and Curves, and Some Mathe-Magical Transformations
Alan R. Jones

Volume IV Learning, Unlearning and Re-learning Curves
Alan R. Jones

Volume V Risk, Opportunity, Uncertainty and Other Random Models
Alan R. Jones

Principles, Process and Practice of Professional Number Juggling

Alan R. Jones

Routledge
Taylor & Francis Group
LONDON AND NEW YORK

First published in paperback 2024

First published 2019
by Routledge
4 Park Square, Milton Park, Abingdon, Oxon OX14 4RN

and by Routledge
605 Third Avenue, New York, NY 10158

Routledge is an imprint of the Taylor & Francis Group, an informa business

Publisher's Note
The publisher has gone to great lengths to ensure the quality of this reprint but points out that some imperfections in the original copies may be apparent.

British Library Cataloguing-in-Publication Data
A catalogue record for this book is available from the British Library

Library of Congress Cataloging-in-Publication Data
Names: Jones, Alan (Alan R.), 1953- author.
Title: Principles, process and practice of professional number juggling / Alan Jones.
Description: Abingdon, Oxon ; New York, NY : Routledge, 2018. | Series:
   Working guides to estimating & forecasting ; volume 1 | Includes
   bibliographical references and index.
Identifiers: LCCN 2017059080 (print) | LCCN 2018004660 (ebook) |
   ISBN 9781315160054 (eBook) | ISBN 9781138063969 (hardback : alk. paper)
Subjects: LCSH: Costs, Industrial—Estimates. | Costs, Industrial—Statistical methods. |
Industrial engineering—Statistical methods.
Classification: LCC TS167 (ebook) | LCC TS167 .J66 2018 (print) | DDC 338.5/1—dc23
LC record available at https://lccn.loc.gov/2017059080

ISBN: 978-1-138-06396-9 (hbk)
ISBN: 978-1-03-283878-6 (pbk)
ISBN: 978-1-315-16005-4 (ebk)

DOI: 10.4324/9781315160054

Typeset in Bembo
by Apex CoVantage, LLC

To my family:
Lynda, Martin, Gareth and Karl
Thank you for your support and forbearance, and for understanding
why I wanted to do this.

My thanks also to my friends and former colleagues at BAE Systems and the wider
Estimating Community for allowing me the opportunity to learn, develop and practice
my profession … and for suffering my brand of humour over the years.
In particular, a special thanks to Tracey C, Mike C, Mick P and Andy L for your
support, encouragement and wise counsel. (You know who you are!)

Contents

Figures

Tables

Foreword to the *Working Guides to Estimating and Forecasting* series

At long last, a book that will support you throughout your career as an estimator and any other career where you need to manipulate, analyse and, more importantly, make decisions using your results. Do not be concerned that the book consists of five volumes as the book is organised into five distinct sections. Whether you are an absolute beginner or an experienced estimator there will be something for you in these books!

Volume One provides the reader with the core underpinning good-practice required when estimating. Many books miss out the need for auditability of your process, clarity of your approach and the techniques you have used. Here, Alan Jones guides you on presenting the basis of your estimate, ensuring you can justify your decisions, evidence these and most of all ensure you keep the focus and understand and focus on the purpose of the model. By the end of this volume you will know how to use, for example, e.g. factors and ratios to support data normalisation and how to evidence qualitative judgement. The next volume then leads you through the realm of probability and statistics. This will be useful for Undergraduate students through to experienced professional engineers. The purpose of Volume Two is to ensure the reader *understands* the techniques they will be using as well as identifying whether the relationships are statistically significant. By the end of this volume you will be able to analyse data, use the appropriate statistical techniques and be able to determine whether a data point is an outlier or not. Alan then leads us into methods to assist us in presenting non-linear relationships as linear relationships. He presents examples and illustrations for single linear relationships to multi-linear dimensions. Here you do need to have a grasp of the mathematics and the examples and key points highlighted throughout the volumes ensure you can. By the end of this volume you will really grasp best-fit lines and curves.

After Volume Three the focus moves to other influences on your estimates. Volume Four brings out the concept of learning curves – as well as unlearning curves! Throughout this volume you will start with the science behind learning curves but unlike other books, you will get the whole picture. What happens across shared projects and learning, what happens if you have a break in production and have to restart learning. This volume covers the breadth of scenarios that may occur and more importantly how to build these into your estimation process. In my view covering the various types of learning and reflecting these back to real life scenarios is the big win. As stated many authors focus on learning curves and assume a certain pattern of behaviour. Alan provides you with options, explains these and guides you on how to use them.

The final volume tackles risk and uncertainty. Naturally Monte-Carlo simulation is introduced and a guide on really understanding what you are doing. One of the real winners here is some clear hints on guidance on good practice and what to avoid doing. To finalise the book, Alan reflects on the future of Manufacturing where this encompasses the whole life cycle. From his background in Aerospace he can demonstrate the need for critical path in design, manufacture and support along with schedule risk. By considering uncertainty in combination with queueing theory, especially in the spares and repairs domain, Alan demonstrates how the build-up of knowledge from the five volumes can be used to estimate and optimise the whole lifecycle costs of a product and combined services.

I have been waiting for this book to be published for a while and I am grateful for all the work Alan has undertaken to provide what I believe to be a seminal piece of work on the mathematical techniques and methods required to become a great cost estimator. My advice would be for every University Library and every cost estimating team (and beyond) to buy this book. It will serve you through your whole career.

Linda Newnes
Professor of Cost Engineering
Department of Mechanical Engineering
University of Bath
BA2 7AY

1 Introduction and objectives

I guess the most inappropriate way to begin a book on estimating is with the words '*I guess…*', but then that's what many people think estimating is all about … guessing, but it's not, there's a lot more to it. (*I have to say that otherwise this series of books will have been a waste of time.*) There is also an impression given by some that 'anyone can estimate', which at its most basic level is true; anyone **can** estimate, and everyone does! We all do it in our daily lives, at work, at home, on the move, but what does it take to be a professional Cost Estimator? The answer to that is Subject Knowledge, Technique, Data… and broad shoulders because there will be plenty of other people, who aren't estimators, who will have an opinion that we've got it wrong … and, yes, sometimes we do. Estimating is not an exact science; it can also be something of a 'dark art' involving an informed judgement.

Also, contrary to popular opinion in some quarters, estimators are not 'mathe-magicians' who can miraculously conjure up an estimate out of nothing. However, the ability to juggle numbers is a pre-requisite skill of an estimator, but out of nothing? That does sound like guessing, or picking a random number! (That said, we can use random numbers to generate an estimate, but there is still a structure to it; we will deal with that under Monte Carlo Simulation in Volume V.)

This series of books aspires to be a practical reference guide to a range of numerical techniques and models that an estimator might wish to consider in analysing historical data in order to forecast the future. Many of the examples and techniques discussed relate to cost estimating in some way, as the term estimator is frequently used synonymously to mean Cost Estimator. However, many of these numerical or quantitative techniques can be applied in other areas other than cost where estimating is required, such as scheduling, or in determining a forecast of physical characteristic, such as weight, length or some other technical parameter.

For a degree of balance, there are some (albeit relatively few) qualitative techniques included to supplement the quantitative or numerical techniques.

This series of volumes does not claim that any or all of these techniques will necessarily give you better answers, but they will be TRACEable, a key theme that we will discuss in Chapter 3.) In fact, the whole series very nearly appeared under the name of

'*The Estimator's Nearly Right Guide to ...*' which in essence sums up the life of an estimator. However, the question of which we should be mindful is whether '*nearly right*' is right enough?

1.1 Why write this book? Who might find it useful? Why five volumes?

1.1.1 Why write this series? Who might find it useful?

The intended audience is quite broad, ranging from the relative 'novice' who is embarking on a career as a professional estimator, to those already seasoned in the science and dark arts of estimating. Somewhere between these two extremes of experience, there will be some who just want to know what tips and techniques they can use, to those who really want to understand the theory of why some things work and other things don't. As a consequence, the style of this book is aimed to attract and provide signposts to both (and all those in between).

This series of books is not just aimed at Cost Estimators (although there is a natural bias there). There may be some useful tips and techniques for other number jugglers, in which we might include other professionals like Engineers or Accountants who estimate but do not consider themselves to be estimators *per se*. Also, in using the term 'estimator', we should not constrain our thinking to those whose Estimate's output currency is cost or hours, but also those who estimate in different 'currencies', such as time and physical dimensions or some other technical characteristics.

Finally, in the process of writing this series of guides, it has been a personal voyage of discovery, cathartic even, reminding me of some of the things I once knew but seem to have forgotten or mislaid somewhere along the way. Also, in researching the content, I have discovered many things that I didn't know and now wish I had known years ago when I started on my career, having fallen into it, rather than chosen it (*does that sound familiar to other estimators?*).

1.1.2 Why five volumes?

There are two reasons:

Size ... there was too much material for the single printed volume that was originally planned ... *and that might have made it too much of a heavy reading, so to speak.* That brings out another point, the attempt at humour will remain around that level throughout.

Cost ... even if it had been produced as a single volume (printed or electronic), the cost may have proved to be prohibitive without a mortgage, and the project would then have been unviable.

So, a decision was made to offer it as a set of five volumes, such that each volume could be purchased and read independently of the others. There is cross-referencing between the volumes, just in case any of us want to dig a little deeper, but by and large the five volumes can be read independently of each other. There is a common Glossary of Terms across the five volumes which covers terminology that is defined and assumed throughout. This was considered to be essential in setting the right context, as there are many different interpretations of some words in common use in estimating circles. Regrettably, there is a lack of common understanding by what these terms mean, so the glossary clarifies what is meant in this series of volumes.

1.2 Features you'll find in this book and others in this series

People's appetites for practical knowledge varies from the 'How do I?' to the 'Why does that work?' This book will attempt to cater for all tastes.

Many text books are written quite formally, using the third person which can give a feeling of remoteness. In this book, the style used is in first person plural, 'we' and 'us'. Hopefully this will give the sense that this is a journey on which we are embarking together, and that you, the reader, are not alone, especially when it gets to the tricky bits! On that point, let's look at some of the features in this series of *Working Guides to Estimating & Forecasting* ...

1.2.1 Chapter context

Perhaps unsurprisingly, each chapter commences with a very short dialogue about what we are trying to achieve or the purpose of that chapter, and sometimes we might include an outline of a scenario or problem we are trying to address.

1.2.2 The lighter side (humour)

There are some who think that an estimator with a sense of humour is an oxymoron. (*Not true, it's what keeps us sane.*) Experience gleaned from developing and delivering training for estimators has highlighted that people learn better if they are enjoying themselves. We will discover little 'asides' here and there, sometimes at random but usually in italics, to try and keep the attention levels up. (*You're not falling asleep already, are you?*) In other cases, the humour, sometimes visual, is used as an *aide memoire*. Those of us who were hoping for a high level of razor-sharp wit, should prepare themselves for a level of disappointment!

1.2.3 Quotations

Here we take the old adage '*A Word to the Wise* ...' and give it a slight twist so that we can draw on the wisdom of those far wiser and more experienced in life than I. We call these little interjections '*A word (or two) from the wise?*' You will spot them easily

A word (or two) from the wise?

'Prediction is very difficult, especially if it's about the future.'
Niels Henrik David Bohr
Danish Physicist and Nobel Laureate
1885–1962

by the rounded shadow boxes. In this one, Niels Bohr (attributed by Ellis, 1970) empathises that the task of estimating (predicting) is not an easy one, especially future events or values. We could take this with a somewhat dismissive *'Why would we want to predict the past?'*, but let's cut the Nobel Laureate a little slack … Estimators are often required to estimate alternative values for past and present events in order to make legitimate comparisons … a process called normalisation, which we will cover in Chapter 6 of this volume. Also, the further into the future, or the further away from our historical base in other dimensions we are, the more challenging the estimating task becomes, and the greater the uncertainty range will be around a single point estimate. The need for a 3-Point Estimate is a recurrent theme through this series of volumes, and is usually expressed as Optimistic, Most Likely and Pessimistic values. We will expand on this in Chapter 4.

1.2.4 Definitions

Estimating is not just about numbers but requires the context of an estimate to be expressed in words. There are some words that have very precise meanings; there are others that mean different things to different people (estimators often fall into this latter group). To avoid confusion, we proffer definitions of key words and phrases so that we have a common understanding within the confines of this series of working guides. Where possible we have highlighted where we think that words may be interpreted differently in some sectors, which regrettably, is all too often. I am under no illusion that back in the safety of the real world we will continue to refer to them as they are understood in those sectors, areas and environments.

For example, in the context of this series of working guides, what do we mean by an 'Estimate'?

Definition 1.1   Estimate

An Estimate for "something" is a numerical expression of the approximate value that might reasonably be expected to occur based on a given context, which is described and is bounded by a number of parameters and assumptions, all of which are pertinent to and necessarily accompany the numerical value provided.

I dare say that some of the definitions given may be controversial with some of us. However, the important point is that they are discussed and considered, and understood in the context of this book, so that everyone accessing these books have the same interpretation; we don't have to agree with the ones given here forevermore – what estimator ever did that? The key point here is that we are able to appreciate that not everyone has the same interpretation of these terms. In some cases, we will defer to the *Oxford English Dictionary* (Stevenson & Waite, 2011) as the arbiter.

1.2.5 Discussions and explanations with a mathematical slant for Formula-philes

These sections are where we define the formulae that underpin many of the techniques in this book. They are boxed off with a header indicative of the dark side to warn off the faint hearted. We will, within reason, provide justification for the definitions and techniques used. For example:

For the Formula-philes: Benford's Law

Consider a collection or system of values in which the leading digit of each constituent value is an integer in range of 1 to 9.
Benford's Law states that the probability of the leading digit being N is given by:

$$pr(N) = \log_{10}\left(1 + \frac{1}{N}\right)$$

1.2.6 Discussions and explanations without a mathematical slant for Formula-phobes

For those less geeky than me, who don't get a buzz from knowing why a formula works (*yes, it's true, there are some estimators like that*), there are the Formula–phobe sections with a suitable less sinister header to give you more of a warm comforting feeling. These are usually wordier with pictorial justifications, and with specific particular examples where it helps the understanding and acceptance.

For the Formula-phobes: One-way logic is like a dead lobster

An analogy I remember coming across reading as a fledgling teenage mathematician, but for which sadly I can no longer recall its creator, relates to the fate of lobsters. It has stuck with me, and I recreate it here with my respects to whoever taught it to me.

(Continued)

Sad though it may be to talk of the untimely death of crustaceans, the truth is that all boiled lobsters are dead! However, we cannot say that the reverse is true – not all dead lobsters have been boiled!

One-way logic is a response to many-to-one relationship in which there are many circumstances that lead to a single outcome, but from that outcome we cannot stipulate what was the circumstance that led to it.

Please note that no real lobsters were harmed in the making of this analogy.

1.2.7 Caveat augur

Based on the fairly well-known warning to shoppers: '*Caveat Emptor*' (let the buyer beware) these call-out sections provide warnings to all soothsayers (or estimators) who try to predict the future, that in some circumstances we many encounter difficulties in using some of the techniques. They should not be considered to be fool-proof or be a panacea to cure all ills.

Caveat augur

These are warnings to the estimator that there are certain limitations, pitfalls or tripwires in the use or interpretation of some of the techniques. We cannot profess to cover every particular aspect, but where they come to mind these gentle warnings are shared

1.2.8 Worked examples

There is a proliferation of examples of the numerical techniques in action. These are often tabular in form to allow us to reproduce the examples in Microsoft Excel (*other spreadsheet tools are available*). Graphical or pictorial figures are also used frequently to draw attention to particular points. The book advocates that we should '**always draw a picture before and after analysing data.**' In some cases, we show situations where a particular technique is unsuitable (i.e. it doesn't work) and try to explain why. Sometimes we learn from our mistakes; nothing and no-one is infallible in the wondrous world of estimating. The tabular examples follow the spirit and intent of Good Practice Spreadsheet Modelling (albeit limited to black and white in the absences of affordable colour printing), the principles and virtues of which are summarised here in Chapter 3.

1.2.9 Useful Microsoft Excel functions and facilities

Embedded in many of the examples are some of the many useful special functions and facilities found within Microsoft Excel (*often, but not always, the estimator's toolset of choice because of its flexibility and accessibility*). Together we explore how we can exploit these functions and features in using the techniques described in this book.

We will always provide the full syntax as we recommend that we avoid allowing Microsoft Excel to use its default settings for certain parameters when they are not specified. This avoids unexpected and unintended results in modelling and improves transparency, an important concept that we discuss in Chapter 3.

Example:

> The **SUMIF**(*range, criteria, sum_range*) function will summate the values in the *sum_range* if the *criteria* in *range* is satisfied, and exclude other values from the sum where the condition is not met. Note that *sum_range* is an optional parameter of the function in Excel; if it is not specified then the *range* will be assumed instead. We recommend that we specify it even if it is the same. This is not because we don't trust Excel, but a person interpreting our model may not be aware that a default has been assumed without our being by their side to explain it.

1.2.10 References to authoritative sources

Every estimate requires a documented Basis of Estimate. In common with that principle, which will be discussed in Chapter 3, every chapter will provide a reference source for researchers, technical authors, writers and those of a curious disposition, where an original, more authoritative, or more detailed source of information can be found on particular aspects or topics.

Note that an Estimate without a Basis of Estimate becomes a random number in the future. On the same basis, without reference to an authoritative source, prior research or empirical observation becomes little more than a spurious unsubstantiated comment.

1.2.11 Chapter reviews

Perhaps not unexpectedly, each chapter summarises the key topics that we will have discussed on our journey. Where appropriate we may draw a conclusion or two just to bring things together, or to draw out a key message that may run throughout the chapter.

1.3 Overview of chapters in this volume

This volume focuses on understanding the key principles, process and practice of professional estimating.

In Chapter 2 we discuss and attempt to clarify the difference and confusion between ambiguous terms like Process, Procedure, Method, Approach and Technique, and how they interrelate.

A robust estimating process is one that supports traceability; in Chapter 3, we define what this means and introduce the theme of TRACEability, and how it supports the all-important Basis of Estimate (BoE). (*An estimate without a BoE is no better than a random number from a TRACEability perspective.*) We go on to discuss how we might score the maturity or robustness of our estimate with an Estimate Maturity Assessment (EMA). The principles of Good Practice Spreadsheet Modelling are also outlined in this chapter, and we introduce the concept of evaluating the Inherent Risks in our Spreadsheets with a qualitative tool called IRiS.

In Chapter 4 we consider the difference and relative relevance of Accuracy and Precision to Estimating and Forecasting, and how we might, or should, interpret these in relation to Primary and Secondary Estimate Drivers. We will discuss why, in general, we want estimates to be accurate but not necessarily precise.

The principle of using Factors, Rates and Ratios in support of an Analogical Method of Estimating is introduced in Chapter 5, and the closely associated and all-important Data Normalisation that we must consider in every Estimates we create, is discussed in Chapter 6. The difference here is that Data Normalisation refers to the principle of ensuring a like-for-like comparison between historical data, whereas the Analogical Method of Estimating uses some of the same basic techniques to create an estimate based on known or perceived differences between what we are trying to estimate and some known historical value. (*I can see some of us looking a bit puzzled; in truth the difference is rather subtle.*)

Estimating is not always about numerical manipulations; estimators need to exercise judgement, often around numerical values, but there will be times when a more qualitative approach may be required. Theoretically, this may then conflict with the principle of TRACEability that we introduce in Chapter 3. In Chapter 7 we introduce some pseudo-quantitative techniques that aid such qualitative judgements, which can then be documented to support TRACEability.

Finally, in Chapter 8, we introduce Benford's Law that is used extensively in fraud detection. It is offered here in the context of assessing the robustness of multiple lower-order estimates from third parties (such as vendor quotations) that we then incorporate into larger project estimates. It may be possible to detect any general 'bias' towards overstated or understated inputs.

1.4 Elsewhere in the 'Working Guide to Estimating & Forecasting' series

Whilst every effort has been made to keep each volume independent of others in the series, this would have been impossible without some major duplication and overlap. Whilst there is

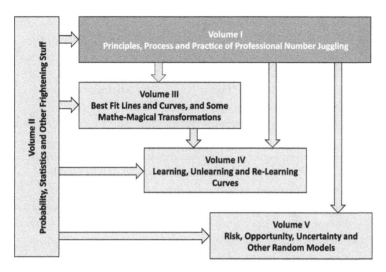

Figure 1.1 Principal Flow of Prior Topic Knowledge Between Volumes

quite a lot of cross-referral to other volumes, this is largely for those of us who want to explore particular topics in more depth. There are some more fundamental potential pre-requisites. For example, the Regression techniques required to calibrate a Learning Curve (i.e. the 'how' and the 'when') are covered in Volume IV, but a thorough understanding and the 'what' and the 'why' assumes we have had access to Volume III. This volume establishes many of the principles of good estimating practice that flow through the other volumes in this series, but also draws on some of the principles of probability and statistics discussed in Volume II.

Figure 1.1 indicates the principal linkages or flows across the five volumes, not all of them.

1.4.1 Volume I: Principles, Process and Practice of Professional Number Juggling

This is where we are now. This section is included here just to make sure that the paragraph numbering aligns with the volume numbers! (*Estimators like structure; it's engrained; we can't help it.*)

We covered this in more detail in Section 1.3, so we will not repeat or summarise it further here.

1.4.2 Volume II: Probability, Statistics and Other Frightening Stuff

Volume II is focused on the Statistical concepts that are exploited through Volumes III to V (and to a lesser extent in Volume I). It is not always necessary to read the associated detail in this volume if you are happy just to accept and use the various concepts, principles and conclusions. However, a general understanding is always better than blind acceptance, and this volume is geared around making these statistical topics more accessible and understandable to those who wish to adventure into the darker art and science of estimating. There are also some useful 'Rules of Thumb' that may be helpful to estimators or other number jugglers that are not directly used by other volumes.

We explore the differences between the different statistics that are collectively referred to as 'Measures of Central Tendency' and why they are referred to as such. In this discussion, we consider four different types of Mean (Arithmetic, Geometric, Harmonic and Quadratic) in addition to Modes and the 'one and only' Median, all of which are, or might be, used by Estimators, sometimes without our conscious awareness.

However, the Measures of Central Tendency only tell us half the story about our data, and we should really understand the extent of scatter around the Measures of Central Tendency that we use; this gives us valuable insight to the sensitivity and robustness of our estimate based on the chosen 'central value'. This is where the 'Measures of Dispersion and Shape' come into their own. These measures include various ways of quantifying the 'average' deviation around the Arithmetic Mean or Median, as well as how we might recognise 'skewness' (where data is asymmetric or lop-sided in its distribution), and where our data exhibits high levels of Excess Kurtosis, which measures how spikey our data scatter is relative to the absolute range of scatter. The greater the Excess Kurtosis, and the more symmetrical our data, then the greater confidence we should have in the Measures of Central Tendency being representative of the majority of our data. Talking of 'confidence' this leads us to explore Confidence Intervals and Quantiles, which are frequently used to describe the robustness of an estimate in quantitative terms.

Extending this further we also explore several probability distributions that may describe the potential variation in the data underpinning our estimates more completely. We consider a number of key properties of each that we can exploit, often as 'Rules of Thumb', but that are often accurate enough without being precise.

Estimating in principle is based on the concept of Correlation, which expresses the extent to which the value of one 'thing' varies with another, the value of which we know or have assumed. This volume considers how we can measure the degree of correlation, what it means and, importantly, what it does not mean! It also looks at the problem of a system of variables that are partially correlated, and how we might impose that relationship in a multi-variate model.

Estimating is not just about making calculations, it requires judgement, not least of which is whether an estimating relationship is credible and supportable, or 'statistically significant'. We discuss the use of Hypothesis Testing to support an informed decision when making these judgement calls. This approach leads naturally onto Tests for 'Outliers'. Knowing when and where we can safely and legitimately exclude what looks like unrepresentative or rogue data from our thoughts is always a tricky dilemma for estimators. We wrap up this volume by exploring several Statistical Tests that allow us to 'Out the Outliers'; be warned however, these various Outlier tests do not always give us the same advice!

1.4.3 Volume III: Best Fit Lines and Curves, and Some Mathe-Magical Transformations

This volume concentrates on fitting the 'Best Fit' Line or Curve through our data, and creating estimates through interpolation or extrapolation and expressing the confidence we have in those estimates based on the degree of scatter around the 'Best Fit' Line or Curve.

We start this volume off quite gently by exploring the properties of a straight line that we can exploit, including perhaps a surprising non-linear property. We follow this by looking at simple data smoothing techniques using a range of 'Moving Measures' and stick a proverbial toe in the undulating waters of exponential smoothing. All these techniques can help us to judge whether we do in fact have an underlying trend that is either linear (straight line) or non-linear (curved).

We begin our exploration of the delights of Least Squares Regression by considering how and why it works with simple straight-line relationships before extending it out into additional 'multi-linear' dimensions with several independent variables, each of which is linearly correlated with our dependent variable that we want to estimate. A very important aspect of formal Regression Analysis is measuring whether the Regression relationship is credible and supportable.

Such is the world of estimating, that many estimating relationships are not linear, but there are three groups of relationships (or functions) that can be converted into linear relationships with a bit of simple mathe-magical transformation. These are Exponential, Logarithmic and Power Functions; some of use will have seen these as different Trend-line types in Microsoft Excel.

We then demonstrate how we can use this mathe-magical transformation to convert a non-linear relationship into a linear one to which we can subsequently exploit the power of Least Squares Regression.

Where we have data that cannot be transformed in to a simple or multi-linear form, we explore the options open to us to find the 'Best Fit' curve, using Least Squares from first principles, and exploiting the power of Microsoft Excel's Solver.

Last, but not least, we look at Time Series Analysis techniques in which we consider a repeating seasonal and/or cyclical variation in our data over time around an underlying trend.

1.4.4 Volume IV: Learning, Unlearning and Re-Learning Curves

Where we have recurring or repeating activities that exhibit a progressive reduction in cost, time or effort we might want to consider Learning Curves, which have been shown empirically to work in many different sectors.

We start our exploration by considering the basic principles of a learning curve and the alternative models that are available, which are almost always based on Crawford's Unit Learning Curve or the original Wright's Cumulative Average Learning Curve. Later in the volume we will discuss the lesser used Time-based Learning

Curves and how they differ from Unit-based Learning Curves. This is followed by a healthy debate on the drivers of learning, and how this gave rise to the Segmentation Approach to Unit Learning.

One of the most difficult scenarios to quantify is the negative impact of breaks in continuity, causing what we might term Unlearning or Forgetting, and subsequent Re-learning. We discuss options for how these can be addressed in a number of ways, including the Segmentation Approach and the Anderlohr Technique.

There is perhaps a misconception that Unit-based Learning means that we can only update our Learning Curve analysis when each successive unit is completed. This is not so, and we show how we can use Equivalent Units Completed to give us an 'early warning indicator' of changes in the underlying Unit-based Learning.

We then turn our attention to shared learning across similar products or variants of a base product through Multi-Variant Learning, before extending the principles of the segmentation technique to a more general transfer of learning across between different products using common business processes.

Although it is perhaps a somewhat tenuous link, this is where we explore the issue of Collaborative Projects in which work is shared between partners, often internationally with workshare being driven by their respective national authority customers based on their investment proportions. This generally adds cost due to duplication of effort and an increase in integration activity. There are a couple of models that may help us to estimate such impacts, one of which bears an uncanny resemblance to a Cumulative Average Learning Curve. (*I said that it was a tenuous link.*)

1.4.5 Volume V: Risk, Opportunity, Uncertainty and Other Random Models

Volume V, the last in the series, begins with a discussion on how we can model research and development, concept demonstration, or design and development tasks when we may only know the objective and not how we are going to achieve it. Possible solutions may be to explore the use of a Norden-Rayleigh Curve, or a Beta, PERT-Beta or even a Triangular Distribution. These repeating patterns of resource effort have been shown empirically to follow the natural pattern of problem discovery and resolution over the life of such 'solution development' projects.

Based fundamentally on the principles of 3-Point Estimates, we discuss how we can use Monte Carlo Simulation to model and analyse Risk, Opportunity and Uncertainty variation. As Monte Carlo Simulation software is generally proprietary in nature, and is often 'under-understood' by its users, we discuss some of the 'do's and don'ts' in the context of Risk, Opportunity and Uncertainty Modelling, not least of which is how and when to apply partial correlation between apparently random events! However, Monte Carlo Simulation is not a technique that is the sole reserve of the Risk Managers and the like; it can also be used to test other assumptions in a more general modelling and estimating sense.

There are other approaches to Risk, Opportunity and Uncertainty Modelling other than Monte Carlo Simulation, and we discuss some of these here. In particular, we discuss the Risk Factoring Technique that is commonly used, and sadly this is often misused to quantify risk contingency budgets.

There is a saying (attributed to Benjamin Franklin) that *'time is money'* and estimators may be tasked with ensuring that their estimates are based on achievable schedules. This links back to Schedule Risk Analysis using Monte Carlo Simulation, but also requires an understanding of the principles (at least) of Critical Path Analysis. We discuss these here and demonstrate that a simple Critical Path can be developed against which we can create a schedule for profiling and to some extent verifying costs.

In the last chapter of this last volume (*ah, sad*) we discuss Queueing Theory (*it just had to be last one, didn't it? I just hope that the wait is worth it*). We show how we might use this in support of achievable solutions where we have random arisings (such as spares or repairs) against which we need to develop a viable estimate.

1.5 Final thoughts and musings on this volume and series

In this chapter, we have outlined the contents of this volume and to some degree the others in this series, and described the key features that have been included to ease our journey through the various techniques and concepts discussed. We have also discussed the broad outline of each chapter of this volume, and reviewed an overview of the other volumes in the series to whet our appetites. We have also highlighted many of the features that are used throughout the five volumes that comprise this series, to guide our journey and hopefully make it less painful or traumatic.

Fundamentally, this volume deals with good practice principles in the science and arts of Estimating.

The trouble with estimating is that it is rarely right from a precision perspective, even when it's not wrong from an accuracy perspective. We could also make that observation equally about other professions ... informed judgement is essential.

However, we must not delude ourselves into thinking that if we follow the techniques in this series of volumes slavishly that we won't still get it wrong some of the time. This will often be because assumptions have changed or were initially misplaced, or we made a judgement call that perhaps we wouldn't have made in hindsight. As Tim Fargo (2014) reassures us, we can and we will make mistakes. From an estimating perspective, it will often be in those assumptions that we or others had to make; it goes with the territory. A recurring theme

A word (or two) from the wise?

'Never give up your right to be wrong, and be sure to give others that right too.'

Tim Fargo
American Author and Entrepreneur

> **TRACE**: Transparent, Repeatable, Appropriate, Credible and Experientially-based

throughout this volume, and others in the series, is that it is essential that we document what we have done and why; TRACEability is paramount. The techniques in this series are here to help guide our judgement through an informed decision-making process, and to remove the need to resort to 'guesswork' as much as possible.

References

Ellis, AK (1970) *Teaching and Learning Elementary Social Studies (3rd Edition)*, Boston, Allyn & Bacon, p.431.

Fargo, T (2014) 'Never give up your right to be wrong, and be sure to give others that right too', Alphabet Success (Blog), Thursday 3rd April, [online] Available from: http://alphabetsuccess.blogspot.co.uk/2014/04/never-give-up-your-right-to-be-wrong.html [Accessed 11–01–2017].

Stevenson, A & Waite, M (Eds), (2011) *Concise Oxford English Dictionary (12th Edition)*, Oxford, Oxford University Press.

<table>
<tr><td>

| 2 |
</td><td>

Methods, approaches, techniques and related terms
</td></tr>
</table>

Every profession has its own language, terms and ways of doing things. Estimating is no different. The trouble is, the terminology used for ostensibly the same thing is not always the same, nor are the same terms always used in a consistent manner. This is due in some part to the choice of terminology used in different industries or in other business processes with which the estimating process necessarily interfaces. It's probably a good idea to define what we mean by certain terms as we discuss the practice or mechanics of compiling an estimate.

2.1 What is the difference between a method, approach and technique?

With regards to the question posed in the heading, it may sound like the cue for a joke, but the question is neither a joke nor is it rhetorical. In terms of common usage, apparently the answer is '*not a lot.*' The terms 'Method', 'Approach' and 'Technique' are used quite extensively, and largely it would seem, interchangeably. This is true also when it comes to estimating. It may be helpful then if we can adopt a common meaning for each term, at least for the purposes of this series of working guides, in order to differentiate between the myriad of different ways that we can create an estimate.

The phrase 'create an estimate' is used here quite deliberately instead of 'compile an estimate' to invoke the sense that estimating is a creative process, rather than merely a calculation process. A structured way of working is very important for an estimator in order to provide a robust audit trail

> ### A word (or two) from the wise?
>
> *'Method is much, technique is much, but inspiration is even more.'*
> **Benjamin N. Cardozo**
> American Lawyer and Associate
> Supreme Court Justice
> 1870–1938

through the calculation stages of that creative process that others can follow, or reproduce, in order to validate or challenge the estimate produced. However, we must not forget that there is that very important element of creativity that cannot be taught or even demonstrated. Ben Cardozo (1931) called this 'inspiration'. We might pick up on Cardozo's profession and add the need to apply 'judgement'; all estimators need a balance of method, technique, inspiration and judgement. Unfortunately, inspiration and judgement and things that come from personal experience and not necessarily from a book, so we will concentrate on the more tangible methods and techniques here.

Other related terms that we will also cover at this stage will be the related terms of 'Process' and 'Procedure', which incorporate or guide our methods and techniques.

Each organisation may choose how it expects estimates to be created based on a number of factors:

- The level of definition available
- The amount and quality of data available
- The time allowed to create the estimate
- The purpose for which the estimate is to be used

Each of these will influence the choice of Approach(s), Method(s) or Technique(s) to be adopted in line with the Estimating Process defined within that organisation, but what do we mean when we use each of these terms?

2.2 Estimating Process

Definition 2.1 Estimating Process

An Estimating Process is a series of mandatory or possibly optional actions or steps taken within an organisation, usually in a defined sequence or order, in order to plan, generate and approve an estimate for a specific business purpose.

An Estimating Process within an organisation exists typically for Quality Assurance purposes to provide a means of assuring that estimates generated by a number of people are done so in a manner that is repeatable and auditable. In some countries, it can be a regulatory requirement if you wish to do business with the government (e.g. US DFARS 252.215–7002: Cost Estimating System Requirements in the USA).

A robust Estimating Process is one that should pass the CLIFF Test:

Closed Loop – Iterative – Flexible – Functional

Closed Loop: Requires feedback from within the process and on completion of the work in question to improve process performance and output accuracy

Iterative: Recognises that requirements get clarified or changed progressively and assumptions evolve. Configuration Control is paramount

Flexible: The process needs to be able to respond appropriately where timescales are short and levels of detail information are lacking; this may require some elements of the process to be optional, conditional or discretionary, whereas other will be mandatory

Functional: Regardless of optional steps omitted, the process must always produce a range of numerical values against an agreed scope of work, and a supporting Basis of Estimate all of which have been challenged or validated and cleared for release

Typically, an Estimating Process would cover the following stages or steps, which fall neatly into the often-cited Initiate-Plan-Do-Review Cycle:

- The means of triggering the estimating process
- The definition and agreement of requirements and assumptions to meet the business purpose in question
- The means of creating, validating or challenging the estimate and the basis of that estimate
- The mechanism by which an estimate is approved by the relevant authority within the organisation for use to meet its business purpose
- The feedback mechanisms that compare the actual outturn with the approved estimate

The Estimating Process would typically require that estimates are created under configuration or version control to ensure that changes in assumptions are agreed and recorded appropriately. Also, typically, they would clarify roles and responsibilities for particular actions or stages of the process.

You will note that the word 'typically' is used quite frequently. This is an acknowledgement that every organisation is different, and will define its estimating process in the context of the environment in which it operates, its management style and culture, and its other key business processes, but usually it would reflect the generic life cycle of an Estimate as depicted in the simplistic generic process in the two examples in Figure 2.1. Each stage of the process may be defined as sub-processes. Note that the Estimate Creation (*the Number Juggling element*) is a small part of the overall Estimating Process.

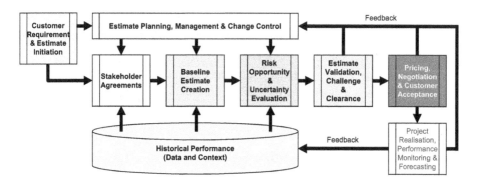

Figure 2.1 Generic Estimating Process

As a consequence, we will not be considering the Estimating Process further in this book other than to acknowledge that some estimating processes may define circumstances where different Estimating Approaches, methods and/or techniques might be applied. Others may not be that prescriptive.

2.3 Estimating Approaches

> ### Definition 2.2   Estimating Approach
>
> An Estimating Approach describes the direction by which the lowest level of detail to be estimated is determined.

Other than implying a sense of direction and a level, the approach to estimating is largely unspecific. We might like to consider it to be the strategic level of describing how we might create an estimate and at what level of detail.

There are three basic modes of approach when it comes to estimating, although most articles on the subject will probably only refer to the two most obvious ones listed here;

the third is more controversial. It is included here in order to highlight a specific point (*rather than merely to give vent to my yearning to be a bit of a rebel*). The three approaches are:

- **Top-down Approach**
- **Bottom-up Approach**
- **Ethereal Approach**

To 'Trekkies' (i.e. fans of Star Trek, the popular science fiction television programme and films), the 'Ethereal' approach must be their favourite, as it is neither Top-down, nor Bottom-up – with this approach, estimates simply materialise out of thin air! Alternatively, this approach might be better labelled the Dramatic Approach in recognition of theatrical instructions 'Enter Stage Left'. More on this later ...

2.3.1 Top-down Approach

In a Top-down Approach to estimating, we would review the overall scope of the task we have been asked to estimate, and identify the major elements of work, resource, cost etc. that characterise the task. We should then ask ourselves whether we have sufficient information to create an estimate at that level for each of these major elements; typically, this would be 'actuals' and some high level technical or programmatic characteristics. If we were not comfortable with the information available we would continue to break the 'challenging' elements of the task down until we reached the level we felt able to create an estimate.

Typically, we might want to consider a natural flow down through the Work Breakdown Structure (WBS), Product Breakdown Structure (PBS) or Service Breakdown Structure (SBS).

Note: We do not have to estimate at the same level across the entire breakdown structure chosen, but we do need to cover all elements of the task at some level. Over time with subsequent iterations of various elements of the estimate, we might want to break down these higher level estimates into more detailed or more refined elements.

An example appears in Figure 2.2.

We would create the top-level estimate simply by aggregating these 'slightly lower' but still high level estimates.

A Top-down Approach is frequently used for creating Rough Order of Magnitude Estimates (RoMs), otherwise known as Ball-Park Estimates, where the level of detail available is limited. As a general rule, we can usually create a top-down estimate in less time than we would do using a Bottom-up Approach. It is this quick turnaround that has perhaps led to the unfortunate reputation that a Top-down Approach is only capable of producing '*quick and dirty*' estimates (i.e. not very accurate or transparent). However, they can legitimately be used in more formal circumstances that require a more definitive estimate such as a customer cost proposal. In fact, the main benefit of working at a higher level is that there is a tendency to use more holistic data from previous projects

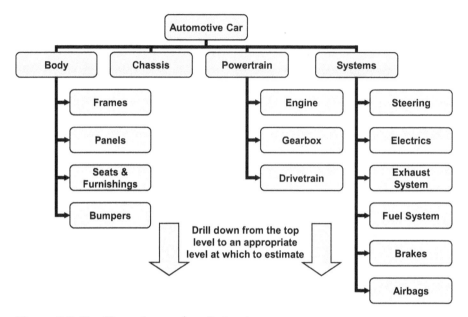

Figure 2.2 Top-Down Approach to Estimating

or products with a reduced chance of some activities or costs falling between the cracks and being missed.

If we think it is appropriate, we can create an estimate at the very highest level, i.e. the total level without any breakdown into major elements. On the other hand, if we break it down too far, we will be running the risk inadvertently of employing a Bottom-up Approach ...

2.3.2 Bottom-up Approach

In a Bottom-up Approach to estimating, we begin by identifying the lowest level that we feel is appropriate to create a range of estimates, based on the task definition available, or that can be inferred. A Bottom-up Approach requires a good definition of the task to be estimated, and is frequently referred to as 'detailed estimating' or as 'engineering build-up'.

The level of detail we select will be influenced by the maturity of the project, product or service, i.e. where we are in the lifecycle, and also the level at which actual data, e.g. costs, are collected in the organisation.

Having created our individual base element estimates, we create the top-level estimate simply by aggregating these lower level estimates.

Bottoms-up estimating – a point of principle!

It is unfortunate that some people refer to it as 'bottoms-up estimating'. This is not an approach with which I am familiar.

Bottoms-up is either a reference to a social drinking salutation, or a submissive deference to a higher authority!

Never produce an estimate 'under the influence' by surrendering your independence!

Figure 2.3 provides an example (abbreviated for simplicity) of what a Bottom-up Approach to estimating might look like.

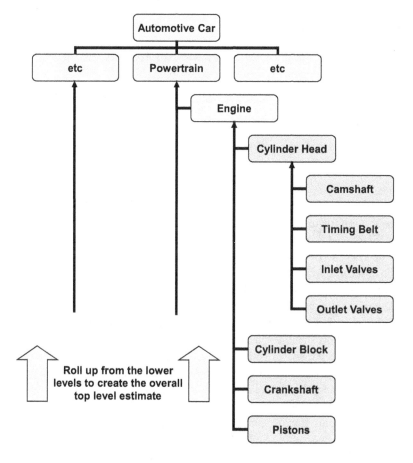

Figure 2.3 Bottom-Up Approach to Estimating

Estimates produced by a Bottom-up Approach are often favoured in some organisations because the levels at which the constituent estimates are compiled are more tangible than those created at a higher level; people can get their minds around the detail of specific tasks. Consequently, bottom-up estimates are often used to gain stakeholder buy-in. However, because the approach focuses on the aggregation of estimates of discrete packages, there is the risk that they miss an element of work. They also require more effort and time to produce than top-down estimates.

In many organisations, bottom-up estimates will be based on the Most Likely values for these low level activities, and as a consequence, it is not unusual for bottom-up estimates to be less in value than the equivalent top-down estimates. However, this is definitely not a suggestion that bottom-up estimates should be inflated arbitrarily. We will discuss this observation further in Volume II Chapter 2 on Measures of Central Tendency, and Volume V Chapter 3 on Monte Carlo Simulation.

2.3.3 Ethereal Approach

Consider an estimate for a discrete element of work, probably as part of a wider scope of work. It is quite legitimate to obtain an estimate by any of the following means:

- Vendor Price Quotation (either by Tender or Single Source)
- Catalogue Price
- Negotiated Price
- Expert Judgement

Whether the Vendor has created the price quotation by either a Top-down or Bottom-up Approach, it is immaterial. To us, receiving that input, we may not know its provenance (*short of asking, and even then, they may not tell us*), so we cannot say legitimately whether it has been created by a Top-down or Bottom-up Approach – it just appears 'out of the blue' into the system.

In relation to values or estimates provided by Subject Matter Experts, we frequently have no idea how they arrive at their esteemed opinion or judgement, and in reality they are probably all different, depending on the expert and the question posed to them. True, they may articulate well how they have arrived at their opinion, but quite often it will be down to 'gut-feel', 'educated guess' or perhaps just a plain 'guesstimate'. If this all sounds a bit hit or miss, consider the alternative. We will probably have decided to ask an expert because we believe that expert has prior knowledge and experience that we do not have, and because we cannot see a viable alternative means of creating an estimate by more scientific or systematic means. If we do not ask the expert, how do we fill the void in our estimate? One possible answer would be 'with our own uneducated guess.' Now, whose guess do we think will be perceived by others to be the more credible?

Whether you accept that an ethereal approach is a legitimate approach in its own right or more of a method or technique (*even though we haven't agreed what we mean by these yet*) within either a Bottom-up or Top-down Approach is for you to decide. A good estimator will listen to the arguments and then form an opinion; we don't all have to agree with the proposition, but hopefully we can agree that there may be values that can enter the estimating process directly, and that we will not know for certain whether they have been created through a Top-down or Bottom-up Approach.

2.4 Estimating Methods

Definition 2.3   Estimating Method

An Estimating Method is a systematic means of creating an estimate, or an element of an estimate. An Estimating Methodology is a set or system of Estimating Methods

There are probably only three basic methods or methodologies by which estimates are created and that we can consider as being **APT**:

*   **Analogy**
*   **Parametric**
*   **'Trusted Source'**

We should recognise that some texts and articles emanating from august learned bodies, and their

> The suffix '-ology' refers to the study or science of the word to which it is attached, so an Estimating Methodology is a study of, or a branch of knowledge relating to Estimating Methods.

seasoned and respected professionals, may refer to other methods. They are not wrong, but appealing to rebel within us, let's consider them instead as either variations to, or a combination of, the three listed here. Some of these 'other methods' are discussed later in this chapter under Section 2.4.4 'Methods that are arguably not methods (in their own right)'.

2.4.1 Analogical or Analogous Method

Firstly, let's consider analogy in general. (*Analogy should not to be confused with 'an allergy', which is an unfortunate, often distressing reaction to some natural or synthetic substances, causing an unpleasant biological or physiological response in a person. ... come to think of it, I have created some estimates that have had a similar response!*)

An analogy can be defined as '*a comparison between one thing and another, made for the purpose of explanation or clarification.*'

Stevenson, Angus & Waite, Maurice (2011) *Concise Oxford English Dictionary (12th Edition)*, definition of Analogy
By Permission of Oxford University Press

For this reason, an Analogical Method is sometimes referred to as the Comparative Method. If we were to draw an analogy between two things we would be making a comparison about certain characteristics they share or that are relatively similar, but also by implication, we would be acknowledging that the two things also have different characteristics because they are only similar, not the same. From an estimating perspective, the recognition of these differences and the degree of similarity or otherwise is essential. Consequently, if we wish to create an estimate by analogy, we need to be able to quantify the degree of similarity and to quantify the extent and impact of the differences on that which we wish to estimate, e.g. weight, cost, schedule time etc.

Definition 2.4 Analogical Estimating Method

The method of estimating by Analogy is a means of creating an estimate by comparing the similarities and/or differences between two things, one of which is used as the reference point against which rational adjustments for differences between the two things are made in order establish an estimate for the other.

By way of example (Table 2.1 and Figure 2.4), consider a new item for which we know the weight and for which we want to create an estimate for the time to assemble it. We search around our database of previous assemblies and find one for which we have both the weight of the item and the actual time to assemble it. Importantly, it is comparable in terms of its 'form, fit and function' to avoid making the mistake of comparing

Table 2.1 Example of a Simple Estimate by Analogy

	Weight (kg)	Assembly Time
Reference Item	25	75
New Item	30	90.0
Ratio	120%	120%

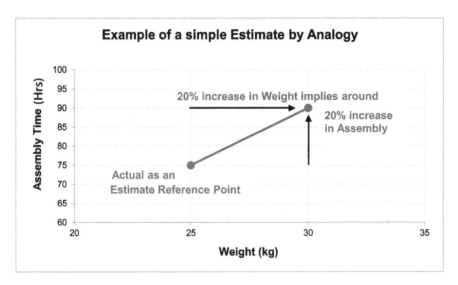

Figure 2.4 Example of a Simple Estimate by Analogy

an apple with an orange, (*not that apples or oranges are assembled*). For instance, it may not be appropriate to compare an assembly where components are joined by a bonding process with an assembly that uses mechanical fasteners. Neither would it be appropriate to compare a structural assembly with that of an electronic box.

Having found a comparable assembly, this becomes our Reference Item. We can now create an estimate of the assembly time for the new item by analogy to the Reference Item, simply by applying the ratio of the weights of the two assemblies to adjust the assembly time of the Reference Item (Ref Item):

$$\text{New Item's Assembly Time} = \text{Ref Item's Assembly Time} \times \frac{\text{New Item's Weight}}{\text{Ref Item's Weight}}$$

It is important that we recognise that the very simplicity of any Analogical Estimating Method implies an underlying Estimating Relationship or metric that passes through the origin, the point (0,0) as depicted in Figure 2.5. The Estimating Relationship or Metric is implied by the slope of the line though the origin, e.g. Man-hours per kilogram, Cost per Non-Destructive Test. Furthermore, this then also suggests that we should be happy to use this implied Estimating Relationship for other such similar tasks. However, we should not assume that the relationship is necessarily valid for other, less similar (e.g. more or less complex) activities.

Figure 2.6 illustrates that a Linear Analogy is a valid approximation to an underlying relationship (e.g. Cost and Cost Driver) for similar tasks or products, but becomes less valid the more dissimilar the task becomes. (Note: Confidence Intervals referred

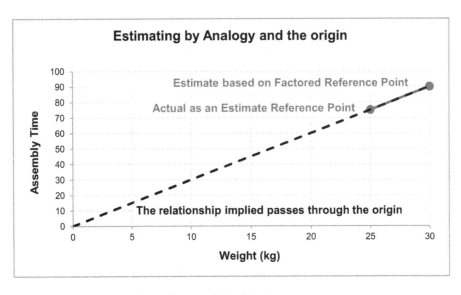

Figure 2.5 Estimating by Analogy and the Origin

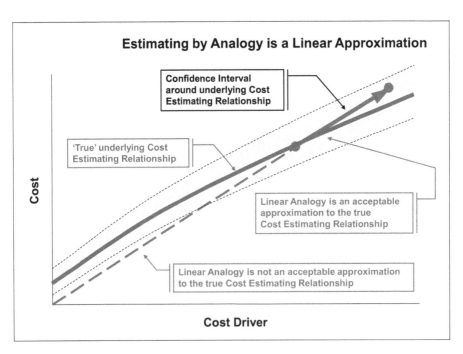

Figure 2.6 Estimating by Analogy is a Linear Approximation

to in the Figure are discussed in Volume II Chapter 3 on Measures of Dispersion and Shape.)

It is not practical to articulate the similarities and differences between two complex products at a high level in minute detail. As estimators we will tend to focus on the primary drivers (see Chapter 4) that are likely to affect the value we wish to estimate, especially where we have elected to use a Top-down Approach. Potentially, where we believe that there is more than one major driver, we will need to consider the interaction between these drivers. For instance, the weights of two products may be good indicators of the relative time to assemble them, but so too might the number of parts to be assembled. Furthermore, we would probably agree that there are circumstances where the number of parts might be an indicator of the weight, and vice versa ... but not necessarily; there might be fewer parts in one, but these parts may be much bigger, depending on the manufacturing methods used. (Note: From a Time to Assemble perspective, a prior sub-assembly of several components would be counted as a single item.) We must ensure that we do not *double count* the differences, or allow different views or dimensions of the differences to cancel each other out without good reason, thereby inadvertently inflating or deflating the estimate inappropriately as a consequence. For that reason, some process and product knowledge are an essential part of the estimator's toolkit, or at least having the 'gumption' or 'nous' to ask the experts in that field.

In the example in Figure 2.7, we have compared two analogies for the same assembly; one based on weight, the other based on parts count. Whilst this gives us a range estimate, we can also consider the combined effect of the two cost drivers.

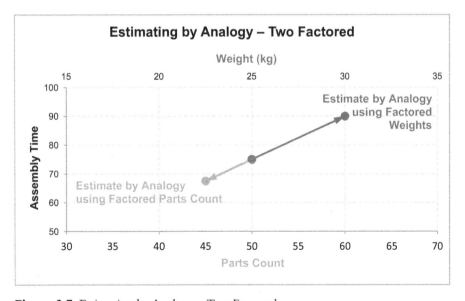

Figure 2.7 Estimating by Analogy – Two Factored

Table 2.2 Estimating by Analogy – Two Factored

	Analogy by Weight		Analogy by Part Count		Analogy by Weight and by Part Count		
	Weight (kg)	Assembly Time	Number of Parts	Assembly Time	Weight (kg)	Number of Parts	Assembly Time
Reference Item Data	25	75	50	75	25	50	75
New Item Data	30		45		30	45	
Ratio (New Item/Ref Item)	120%	120%	90%	90%	120%	90%	108%
New Item Time Estimate		90.0		67.5			81

If we simply take the easy option of averaging the two views, we run the risk of cancelling out some of the true underlying difference. It also gives us a completely different answer to applying both the analogies sequentially, as illustrated in Table 2.2. These different permutations are considered more fully in Chapter 5 on Factors, Rates and Ratios.

However, enough, I am beginning to ramble now, and am straying into 'technique' territory. These will be discussed further in Chapter 5.

To test the sensitivity of the estimate we have created, we can do one of two things:

- Flex our assumptions in terms of the degree of difference
- Make one or more other analogies using different reference points (see Figure 2.8), at which juncture we may want to start to look at a parametric method instead

---

Caveat augur

As we will see in the discussion below on Parametric Methods, it can be argued that the general principle of estimating by analogy using a single driver is pessimistic when scaling up, and optimistic when scaling down with a linear model. The use of additional drivers is a means by which we can mitigate these extremes

---

Some of us may be feeling uncomfortable with the realisation that an Analogical Method usually implies a linear relationship that passes through the origin. However, it doesn't have to be the case. In Volume III Chapter 5 we will be discussing some common non-linear relationships that reside in the real world of the estimator. These are Exponential, Logarithmic and Power functions, all of which are curves that can be transformed into a linear format by a simple bit of 'mathe-magical' know-how (it's the equivalent of a conjuring trick really).

We can then apply the principles of Linear Analogy to these modified transformed curves and then transform the results back into real number values.

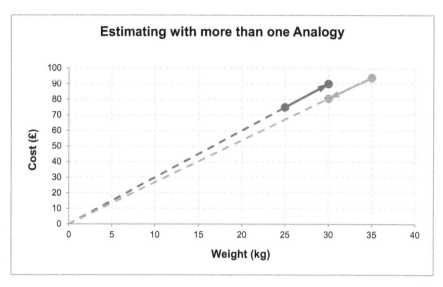

Figure 2.8 Estimating with More Than One Analogy

2.4.2 Parametric Method

If we were to look at the various dictionary definitions of 'Parameter' available to us online, we might start to scratch our heads and feel a little confused. Unfortunately, some of them refer to it as a mathematical or statistical term and talk about variables in an equation that are constant, or constants in an equation that vary. Quite correctly, you would probably conclude '*that's an oxymoron, if ever there was one*' (and regrettably, it may help reaffirm that preconceived idea you may already have had that mathematicians and statisticians are from a different planet!).

Fortunately, some sources are much clearer on the definition:

1. *Measurable or quantifiable characteristic of a system*
2. *[Mathematics] A quantity which is fixed for the case in question but may vary in other cases*
3. *A limit or boundary which defines the scope of a particular process or activity*

Stevenson, Angus & Waite, Maurice (2011) *Concise Oxford
English Dictionary (12th Edition)*, definition of Parameter
By Permission of Oxford University Press

It is the second definition that appears to cause the difficulty. The word 'fixed' is often used synonymously with the word 'constant', but a constant is absolute, whereas we may elect to hold a parameter at a fixed value – its value is held constant conditionally, i.e. until the conditions change. Interpreting this philosophical debate, a Parametric Estimating

Method is one which uses parameters (or variables) to describe the relationship between the '*thing*' we want to estimate as a function of other physical, technical, programmatic, resource or performance characteristics (etc.) of that '*thing*' based on an analysis of other '*things*' with similar characteristics. The '*thing*' might be a project, product or service.

Definition 2.5 Parametric Estimating Method

A Parametric Estimating Method is a systematic means of establishing and exploiting a pattern of behaviour between the variable that we want to estimate, and some other independent variable or set of variables or characteristics that have an influence on its value.

The fundamental difference between parametric estimating and analogical estimating is that parametric estimating is based on an assessment of more than one cost reference point, either at the time that the estimate is being created, or based on a general form of relationship that has been determined previously. Importantly, as the estimate will be based on a number of past actual observations, it should be possible to draw some statistical inferences on the robustness (or otherwise) of the estimate produced.

Figure 2.9 provides an example of this 'pattern of behaviour' by creating a scatter plot of previous actual assembly times for a range of products of a similar type against the weight of the finished product, and looking for the Best-Fit Line or Curve through the data.

An example of utilising an existing parametric relationship could be a 'rule of thumb' relationship such as the three-quarter power rule used in Allometrics for determining the Basal Metabolic Rate based on an animal's weight or mass (Kleiber's Law, 1932)

> **Allometry:** The study of a creature's body size in relation to its shape, anatomy, physiology, maturity and behaviour

If we accept this historical pattern of behaviour we only need a single cost reference point in order to create an estimate for a new ship by factoring the cost in proportion to the ratio of the weights raised to the power of two-thirds. Although we have only used a single cost reference point directly, this does not mean that we have used an Analogical Method, because we have assumed the pattern of behaviour established previously, expressed as a formula. We have indirectly adopted the multiple cost reference points that created the two-thirds power rule in the first place. A Parametric Method is sometimes referred to as a Formulaic Method.

In the petro-chemical industry estimators may use Chilton's Law (Turré, 2006) which is another power rule. The general rule is the six-tenths or 60% power rule and relates cost to size, but in specific cases of different sub-systems or equipment, depending on their

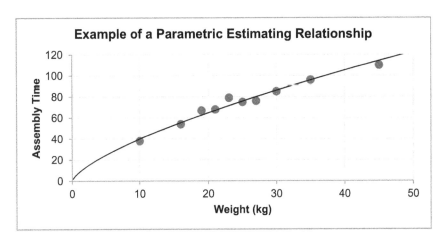

Figure 2.9 Example of a Parametric Estimating Relationship

level of complexity, the power is cited to vary between 0.35 and 0.85 (*we should say 'allegedly' here to keep the lawyers happy – there will always be exceptions*) with values

Chilton's Law: $Cost = Constant \times Size^{0.6}$

closer to one for more complex products. So, if we are to accept the principles of Chilton's Law and Kleiber's Law, then we can scale an estimate parametrically based on the ratio of size raised to a power.

The similarity between these two empirical laws (and there are other equivalent ones) and the examples discussed under Analogical Methods is quite marked. The implication is stark, in that an analogical comparative estimate assumes a default power rule of one, or 100%, which is the value that Chilton 'reserved' for the most complex products or systems. This would tend to suggest that as the power rule 'proportion' usually takes a value of less than one then estimating by analogy will always create a greater value when scaling upwards, and a smaller value if scaling downwards (see Table 2.3):

Table 2.3 Estimating by Analogy Pessimism/Optimism in Relation to a Parametric Power Rule

	Size Ratio	Parametric Relationship	Analogical Comparison	
	Power Rule	0.7	1	
Size Ratio	1.25	1.17	1.25	Analogy > Parametric
	1	1.00	1.00	
	0.8	0.86	0.80	Analogy < Parametric

> **Note:** We can make two separate analogies to give us a level of confidence, but that does not make it parametric *per se* as we would need to understand the potential sensitivity in the adjustments we have made. However, if we infer that the two base comparators are linked and that we can rationalise the difference between them by appropriate adjustments of the key drivers, then we have the making of a parametric relationship.

2.4.3 'Trusted Source' Method

The third method in common use for establishing an estimate is to use someone else's values! (*Thoughts of 'sloping shoulders' come to mind.*) There are, of course certain cases where we might want to use this method as depicted in Table 2.4.

In some circles, when Expert Judgement is used, this can be referred to (somewhat derogatively) as a 'Guessed-imate', which on one level is understandable.

You would be correct in thinking that the 'Trusted Source Method' is very closely linked with an Ethereal Approach, but it is not an exclusive alignment. For a complex estimate with a number of constituent elements, the basic approach of which may be Top-down or Bottom-up, there is nothing to stop us including 'Trusted Source' estimates for discrete elements within the overall estimate.

The Trusted Source Method does not prevent the Estimator seeking or performing an analysis of inputs from several such sources, such as a Supplier Tender Competition.

... a word (or two) from the wise?

"... there was method in his madness"
William Shakespeare
Hamlet
c.1601

Why call it the 'Trusted Source' Method?

Let's put it the other way round: why would you go to a source you didn't trust?

Oh, because there was no other option available! Well then, perhaps we should consider a fourth method – the Method of Last Resort.

Surely, in this case Shakespeare would have added '... but there was also madness in his method.'

If you don't trust it, don't use it!

2.4.4 Methods that are arguably not methods (in their own right)

We should recognise that some texts and articles, emanating from august learned bodies and from seasoned and respected professionals, may refer to other methods such as

Table 2.4 Situations which may Require the 'Trusted Source Method'

Situation	Trusted Source
Cost Estimates for low value commodity items	Catalogue Item Prices
Bought Out Proprietary Items	Vendor Quotations
Tasks we perform very infrequently or where we have no specific data	Expert Judgement
Where buy-in is required from those responsible for project or product delivery or contract execution	Internal Stakeholders

'Engineering Build-up' and 'Extrapolation from Actuals' but in reality these are really just combinations of the three basic methods that we have already discussed:

Engineering Build-up

By the definitions proposed earlier in this chapter, this would fall into the category of a Bottom-up Approach. The methods used within Engineering Build-up would still be Analogy, Parametric and Trusted Source, but at a lower level of detail than one might use with a Top-down Approach.

Extrapolation from Actuals

The principle of extrapolation is to project forward beyond the current data. Fundamentally there are two main methods at our disposal. We might analyse the trend in the actual to date in relation to the achievement, and continue it forward to some future value. This sounds very much like we are identifying a pattern of behaviour and therefore we would be using a parametric method. Alternatively, we could take a more simplistic view that the current position is representative of the future. By analogy we are saying that the future cumulative performance is similar to the current cumulative performance, and factor accordingly.

Simulation

There is an argument that Simulation is a Methodology in its own right. However, on the basis that many simulations utilise data extracted from statistical distributions, which are themselves an arrangement of multiple data points with a defined pattern of behaviour, this then allows us to generate a range of possible outcomes. For that reason, Simulation is considered to be a parametric technique within this series of guides.

Furthermore, we can use Simulation as a technique to estimate the time taken for a discrete task and then use this by Analogy to create an estimate for the 'real task' in question.

We can do this by pretending to carry out a small task (e.g. making a sandwich, drilling a hole) by 'going through the motions', i.e. pretending that we are doing the task, and timing how long we take.

2.5 Estimating Techniques

Definition 2.6   Estimating Technique

An Estimating Technique is a series of actions or steps conducted in an efficient manner to achieve a specific purpose as part of a wider Estimating Method. Techniques can be qualitative as well as quantitative.

Techniques are generally used within a chosen Estimating Method but are not necessarily constrained to any particular approach. A method will consist of one or more techniques. The majority of the numerical techniques are by definition parametric in nature as they are looking for a pattern of behaviour in the data, or they are assuming some other prior pattern established or observed historically. However, there are some simplistic numerical techniques that are appropriate to use within an analogical method. By its very nature there are few (*if any*) numerical techniques associated with the Trusted Source Method, although there may be some more qualitative techniques that might be used, such as the Delphi Technique to gain a consensus or to narrow the range of opinions from a group of Subject Matter Experts.

The detail of the estimating techniques is the main focus of this book. Most of the techniques discussed here are purely numerical in nature, but there are others which are more pseudo-quantitative, and it can be argued that they are really qualitative techniques that have been 'numericalised'. Some of the techniques discussed here are in relation to analysing data as an enabler to estimating; other techniques can be considered to be both analytical and predictive.

2.6 Estimating Procedures

Definition 2.7   Estimating Procedure

An Estimating Procedure is a series of steps conducted in a certain manner and sequence to optimise the output of an Estimating Approach, Method and/or Technique.

We will only discuss procedures in this book where we see that a prescribed sequence of steps would help us get the benefits of using any technique more quickly or more holistically. That said, it is generally recognised that there is a natural sequence to preparing and analysing data prior to estimating that promotes good practice. This generally accepted good practice has been adapted here (Figure 2.10) to reflect where we might want to consider our choice of Estimating Approaches, Methods and Techniques.

Some may consider this to be a process because of the flowchart illustration, but the main difference is that a process includes some definition or assignment of roles and responsibilities in addition to a natural sequence of activities.

The choice of Approach and Method will influence our search for historical data, but the availability of data will also constrain our choice of Approach and Method. Perhaps the best advice is to be optimistic (*having a glass is half full attitude*) and assume that the

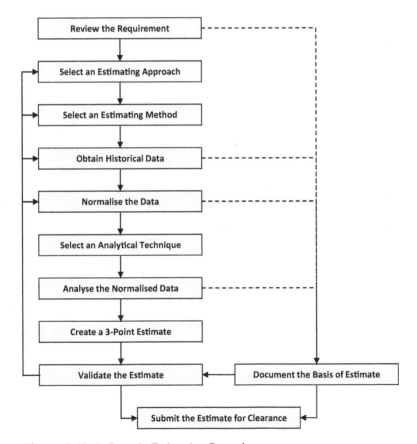

Figure 2.10 A Generic Estimating Procedure

data is there to be harvested, not the opposite (*i.e. the glass is half empty*) and fail to look because it is expected that there is nothing to find (*a somewhat self-fulfilling prophecy*).

The penultimate step before submitting the estimate for clearance is to 'Document the Basis of Estimate', but there is a strong argument that this is better performed progressively in parallel to the other steps in the procedure to assure its quality and to mitigate against recording omissions. Consequently, this step would then become 'Review and Confirm the Basis of Estimate'. Good Practice in Spreadsheet Modelling encourages this also. We will discuss both these topics further in Chapter 3.

2.7 Combining Approaches and Methods

In the main it is fair to consider that we can combine any Estimating Approach with any Estimating Method. The only exception to this 'rule of thumb' is that the only Method that can be used logically with an Ethereal Approach is that of 'Trusted Source' (Table 2.5).

When it comes to the numerical techniques discussed within this book most of them can be used with Top-down and Bottom-up Approaches, and the majority of them support a Parametric Method, but not exclusively; some of them can be used as part of an Analogical Method.

To be fully effective, the analogical method requires an estimator to have a good product or domain knowledge, or to be able to ask those insightful (*not inciteful!*) questions of those who have that domain knowledge in order to interpret the potential cost drivers, their impact and their interaction. In contrast, whilst product or domain knowledge is still important, a Parametric Method needs an estimator to possess good data analysis skills in order to be able to interpret the significance of the potential cost drivers. However, as its title suggests, the 'Trusted Source' Method implies that the product or domain knowledge and understanding of the potential cost drivers is vested largely, if not exclusively, in the 'Trusted Source'.

However, in generating an estimate by more than one Approach, Method or Technique, there is a risk that they shake our confidence in what we might have thought was a good estimate just as easily as boost our confidence in one, as illustrated in Figure 2.11. In many cases we may find that there are clear, but sometimes subtle, differences in the interpretation of the scope implicit in the different Bases of Estimate. We should always

Table 2.5 Valid Combinations of Estimating Approaches and Methods

Valid combinations of Estimating Approaches and Methods		Estimating Method		
		Analogy	Parametric	Trusted Source
Estimating Approach	Top-down	✔	✔	✔
	Bottom-up	✔	✔	✔
	Ethereal	✘	✘	✔

Caveat augur

No Estimating Method, Approach or Technique is infallible. It is always better to create an estimate via more than one method, approach and/or technique and to compare the results obtained than to rely on a single potential answer.

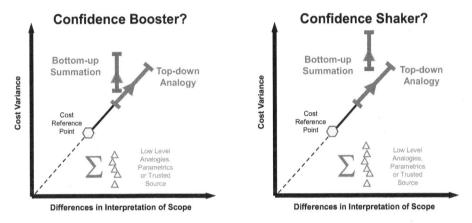

Figure 2.11 Multiple Estimates Using Different Approaches, Methods and Techniques

test the sensitivity of any Estimate to take account of differences in key assumptions. *One thing worse than an over-confident estimator is a deluded one!*

2.7.1 Choice of Estimating Approach for a chosen Estimating Element

For each element of the estimate we need to plan how we are going to create it and validate it. All professional estimators can exercise judgement on the way they create their estimates. We do not have to do every element in the same way. The decision flowchart in Figure 2.12 is not meant to be a definitive framework, but it serves to illustrate some of the thought processes to consider when choosing an appropriate Estimating Approach. That does not exclude other or alternative decisions being made. For instance, just because for a particular element of the estimate we have good definition of the task required, and we are not being rushed for a response to create a Rough Order of Magnitude Estimate, it does not mean we cannot select an Ethereal Approach … but it may not be the best approach if we want to have a complete audit trail; a Top-down Approach may be a better option.

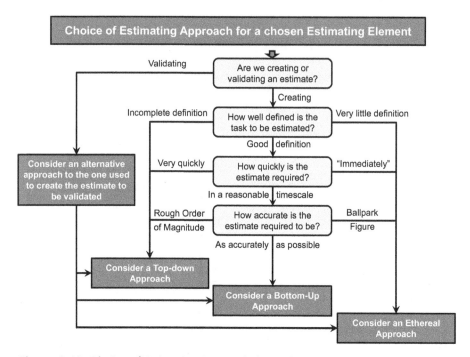

Figure 2.12 Choice of Estimating Approach for a Chosen Estimating Element

Selecting one approach now does not mean we can't change it later once we have reviewed the method and techniques open to us.

2.7.2 Choice of Estimating Method for a chosen Estimating Approach

Having selected a preferred approach, we should now decide on an Estimating Method. As for the Estimating Approach, we also have a choice in respect of the Estimating Method.

Some of the decisions we make will be influenced by the availability of data and in the prior choice of Approach. For instance, it does not make sense to choose a Parametric or Analogical Method if we have already decided on an Ethereal Approach.

Figure 2.13 highlights some of the options we can consider. Again, having multiple data points does not preclude us from using an Analogical or Comparative Method based on just one cost reference point, although we should be asking ourselves why we are choosing to ignore the other data points available to us.

We will be discussing Estimate Drivers in Chapter 4.

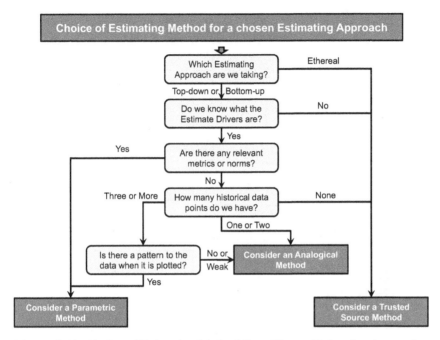

Figure 2.13 Choice of Estimating Method for a Chosen Estimating Approach

2.7.3 Choice of Estimating Technique for a chosen Estimating Method

This is where the choices begin to open up, and a single all-encompassing Flow Diagram would then begin to resemble the tracks of a mainline railway station hub viewed from a high level, with lines criss-crossing all over the place, and with text that would be too small to read. Instead, we'll depict it as a series of four Flow Diagrams covering Techniques that can be used with an Analogical Method, Parametric Methods for time-based and non-time-based data and guidance on choosing when a Simulation Technique might be appropriate (Figures 2.14 to 2.17). These diagrams not intended to be definitive in any absolute sense but are offered as a guide to help steer us in an appropriate direction; other scenarios may arise that these don't cover but the basic thought processes and logic may still be helpful in stimulating ideas we discuss later in the book.

2.8 Chapter review

In this chapter we looked at potential differences between a range of different terms, many of which are used interchangeably both in society at large, and also within the

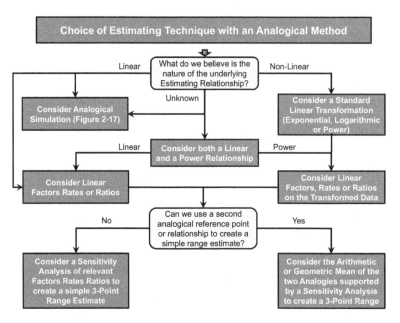

Figure 2.14 Choice of Estimating Technique with an Analogical Estimating Method

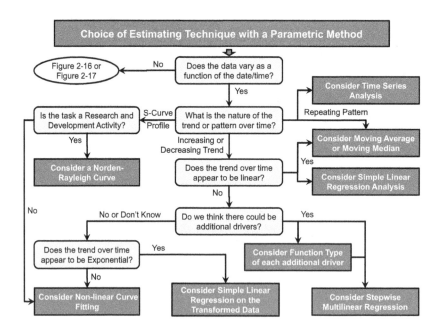

Figure 2.15 Choice of Estimating Technique with a Parametric Estimating Method (1)

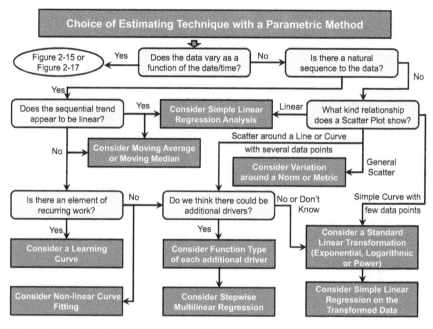

Figure 2.16 Choice of Estimating Technique with a Parametric Estimating Method (2)

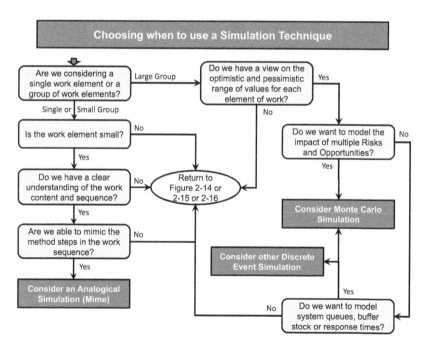

Figure 2.17 Choosing when to use a Simulation Technique

estimating community. In that sense perhaps, with eight definitions, this has been something of a '*defining moment*' early on in the series of books *(sorry, but the standard of humour is really not going to improve)* but there will be other definitions introduced progressively as and when it is appropriate.

In discussing Estimating Approaches and Estimating Methods, we have been a little controversial in defining an 'Ethereal Approach' and 'Trusted Source Method', but whilst there is an element of '*tongue-in-cheek*' in terms of the phrasing used here, they are evocative of the serious point being made in that estimating is not just about juggling numbers and performing calculations; it requires judgement and interpretation ... and of course knowing who we can or cannot trust to provide a sensible, balanced and honest answer.

The main thing is not to be too pretentious about the precise definitions (*we'll define what we may mean by precision in Chapter 4, by the way*) but to 'go with the flow'.

A process guides or instructs us on the who, what, why, where, when and how we should create an estimate. The process should tell us what we must do (mandatory) and what we may do (discretionary). The 'how' may require us to follow a specific procedure or work instruction, or it may merely advise or guide us to perform specific activities in a set sequence. Typically we would perform the analytical activity of the 'how' through an appropriate technique for a chosen method and approach.

References

Cardozo, BN (1931) *Law and Literature and Other Essays and Addresses*, New York, Harcourt Brace & Co. p.163.

Kleiber, M (1932) 'Body size and metabolism', *Hilgardia*, Volume 6: pp.315–351.

Stevenson, A & Waite, M (Eds) (2011) *Concise Oxford English Dictionary (12th Edition)*, Oxford, Oxford University Press.

Turré, G (2006) 'Plant capacity and load' in Foussier, P, *Product Description to Cost: A Practical Approach, Volume 1: The Parametric Approach*, London, Springer-Verlag pp.141–143.

3 Estimate TRACEability and health checks

The Estimating Process is a key business process for many, if not most, organisations. As a key business process, there is an implied requirement to meet the organisation's Quality Management expectations, one of which is that the process must be auditable. The Estimating Process's means of providing such an audit trail includes a well-documented Basis of Estimate, and a clear easy-to-follow set of calculations. A good Basis of Estimate also tells us the source of the data and assumptions used.

Was Pavlov looking at the psychology of estimating here, encouraging us not to accept data blindly but to trace from where it has come in order to understand its limitations?

3.1 Basis of Estimate, TRACEability and estimate maturity

As we have already stated an Estimate without a Basis of Estimate might as well be a random number. The Basis of Estimate provides the context that bounds the validity of the Estimate's numerical value. So what do we mean by a 'Basis of Estimate'?

Definition 3.1 Basis of Estimate (BoE)

A Basis of Estimate is a series of statements that define the assumptions, dependencies and exclusions that bound the scope and validity of an estimate.

It provides an audit trail between the Estimating Plan (articulated through the elements of ADORE) and the Estimate Value created, through a series of logical structured statements that record the approach, method and potentially techniques used, as well as the source and value of key input variables and parameters, and as such supports the objectives of Estimate TRACEability

In order to make sense of this definition, we should explore what we have somewhat glibly referred to as '*assumptions, dependencies and exclusions*'.

*   An **Assumption** is something that we take to be true or expect to come to fruition in the context of the estimate (*although it may not actually be the case*). It may be better if we recognised that an assumption is something that we take to be *broadly* true. (*Nothing spoils a good estimate more than a bad set of assumptions.*)
*   A **Dependency** is something to which an estimate is tied, usually an uncertain event outside of our control or influence, which if it were not to occur, would potentially render the estimated value invalid. If it is an internal dependency, the estimate and schedule should reflect this relationship. However, it is the external dependencies that must be clearly and correctly identified as these often are critical in ensuring that an estimate is viable or not.
*   An **Exclusion** is condition or set of circumstances that we have designated to be out of scope of the current estimating activities and their output (*no real surprise there then!*). However, it is only practical to document those exclusions in the BoE which it might have been reasonable for others to expect had been included in the estimate (*not the oxymoron that it at first sounds*).

It is recognised that for some organisations it may be considered appropriate to list risks and opportunities included or excluded in the Basis of Estimate. This is largely a case of organisational preference or potentially a contractual requirement. We will cover risks and opportunities in Volume V in the main, but will touch on the topic elsewhere as our discussions evolve.

A Basis of Estimate is that which differentiates an estimate from being just 'any other number'. Furthermore, a good BoE is TRACEable – Transparent, Repeatable, Appropriate, Credible and Experientially-based:

*   Transparent – it is clear and unambiguous with nothing hidden
*   Repeatable – another estimator could reproduce the same results with the same information
*   Appropriate – it is justifiable and relevant in the context it is used
*   Credible – it is based on reality, or a pragmatic reasoned argument that can be understood and is believable
*   Experientially-based – it can be underpinned by reference to recorded data (evidence), or prior confirmed experience or expertise

Based on what we have already discussed, this applies to all estimates created by an Analogical or Parametric Method. Ethereal 'Trusted Source' based estimates are not fully TRACEable; with the Transparency being more of a 'Translucency', and as such it should be the method and approach of last resort!

There is also an implication that these principles of TRACEability apply equally to any Estimating Model or Spreadsheet used to generate, compile or process the numerical values. We will discuss what 'Good Practice' looks like for Spreadsheets in Section 3.3.

3.1.1 Building bridges between two estimates

Not to be confused with 'two estimates for building bridges'!

A good Basis of Estimate allows us to understand and demonstrate that we have indeed met the principles of TRACEability. It should also allow us to bridge seamlessly between two versions of an estimate so that any incremental changes as illustrated in Figure 3.1 also meet the principles of TRACEability.

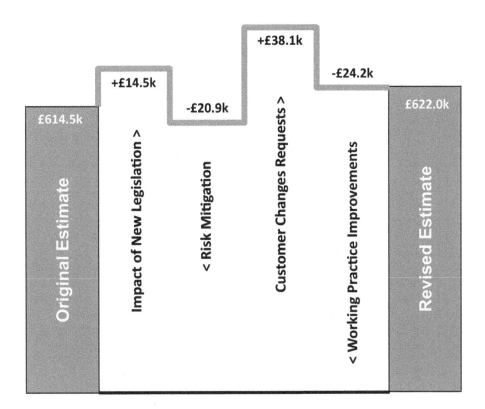

Figure 3.1 Bridging Between Two Estimates

3.2 Estimate and Schedule Maturity Assessments (or health checks)

3.2.1 Estimate Maturity Assessment (EMA)

Within some organisations there is sometimes a culture that accepts an estimator's view of what may be considered to be a 'good number' within the context and caveats defined in a Basis of Estimate, but then going forward, those organisations sometimes appear to focus solely on the 'number', losing or inadvertently disregarding the Basis of Estimate (*and which estimators amongst us have not been there, or seen that happen?*). The reason for this perhaps is that it is easy to remember a number rather than pages of supporting notes and evidence.

The corollary to this behaviour is another, somewhat reactive behaviour that people who are asked to commit to a budgetary value, especially where the task is not fully defined, are prone to inflating the requirement (*it's called protecting their backsides, or some other similar vernacular!*). By implication the culture of such an organisation becomes one in which the term '*budget*' is taken as a pseudonym for '*not to exceed*', and may percolate through the supply chain internally and externally.

Recognising the risk of these behaviours, BAE Systems (Smith, 2013) developed an Estimate Maturity Assessment framework or EMA (*usually pronounced as 'Emma'*) that allows estimators and others to grade the maturity of the Basis of Estimate against a defined framework. The EMA Framework provides a 'score' of 1 to 9 against which a Basis of Estimate can be subjectively graded, thereby providing that all-important 'Health Warning' on an estimate.

In Table 3.1 we summarise the nine EMA Levels:

Table 3.1 Estimate Maturity Assessment Levels

Level	Estimate based on …
EMA9	Precise definition with recorded costs of the exact same nature to the Estimate required
EMA8	Precise definition with recorded costs for a well-defined similar task to the Estimate required
EMA7	Precise definition with validated metrics for a similar task to the Estimate required
EMA6	Good definition with metrics for a defined task similar to the Estimate required
EMA5	Good definition with historical information comparison for a defined task similar to the Estimate required
EMA4	Defined scope with good historical information comparison to the Estimate required
EMA3	Defined scope with poor historical data comparison to the Estimate required
EMA2	Poorly defined scope with poor historical data comparison to the Estimate required
EMA1	Poorly defined scope with no historical data comparison to the Estimate required

Source: © 2012 BAE Systems. Reproduced by kind permission of BAE Systems

If you are thinking that the EMAs bear an uncanny resemblance to Technology Readiness Levels (TRLs) then you would be right as these provided the inspiration to my colleague Frank Berry

at the time, who conceived the EMAs. If Frank was the 'father' of EMA, then another colleague, Ed Smith, must have been the surrogate 'mother', nurturing EMA into a fully developed tool within the business … no offence, Ed, you did a great job!

The simplicity of the EMA is the key to its strength and ultimately to its success:

- Largely independent of the value of the estimate, it concentrates only on the methods and techniques used to develop the estimate
- An EMA allows a 'Health Warning' to be applied to an Estimate so that any future user can be alerted to the robustness or otherwise of the estimate
- The best person to assign an EMA rating is the estimator who developed the estimate – after all, they should know the relative strengths and weaknesses of what they have done
- An Independent EMA can be used as part of a peer review process to provide a measure of process assurance
- An EMA can be assigned to any constituent element of a Bottom-up Estimate. In order to roll-up the constituent's elements into a top-level EMA, the individual EMA scores are weighted by the relative proportion of the constituent estimates to the whole. (*This is why we say that the EMA is only 'largely independent' and not wholly independent of the value*)
- It supports the concept of an iterative approach to estimating, encouraging estimators to focus on the higher value, lower EMA elements to improve the overall robustness of the total estimate
- It can be used as an Estimate Planning aid: by understanding what our expectations are of the likely Maturity of our Estimates, based on the information available, we can elect to concentrate on resolving the less mature, higher value elements of the estimate

Table 3.2 provides an example of a Weighted Average Bottom-Up EMA.

Based on the rationale that any score of 7 and above should be quite 'comforting' in that it shows that the estimate has been based on mature information, we would probably not feel the need to revisit those elements unless the requirement changed dramatically. However, even though in principle we would feel a little uncomfortable with scores in the 1 to 3 region, all the values in this example are for relatively low value work packages; it would probably not be the best use of our time trying to perfect an element of relative insignificance to get a more robust estimate overall.

The element that should cause us the greatest degree of concern is Work Package 4; it is the second highest by value and is scoring an EMA Level 4. Perhaps this is where we should concentrate our future efforts?

Getting middle range values overall should not be unexpected. If we are consistently getting scores at the higher end, then perhaps we should be asking ourselves:

- '*Are we really that good?*'
- '*Are we interpreting our understanding of the Basis of Estimate correctly?*'
- '*Are all our product/services in one or other of the two extremes of Task Familiarity?*'

Table 3.2 Example of a Weighted Average Bottom-Up Estimate Maturity Assessment

Estimating Element	% of Total by Value	Descending Value Rank	EMA Score	Contribution to Overall EMA	Descending EMA Rank
Work Package 1	7%	6	7	0.49	6
Work Package 2	10%	5	6	0.6	5
Work Package 3	1%	10	1	0.01	10
Work Package 4	19%	2	4	0.76	4
Work Package 5	6%	7	3	0.18	7
Work Package 6	2%	9	2	0.04	9
Work Package 7	3%	8	5	0.15	8
Work Package 8	12%	4	8	0.96	3
Work Package 9	23%	1	7	1.61	1
Work Package 10	17%	3	6	1.02	2
Total	**100%**			**5.82**	

In broad terms, we can align our expectations of likely EMA scores to the Product/Service Familiarity as shown in Figure 3.2. Task Familiarity is discussed in more detail in Chapter 6 on Normalisation.

Caveat augur

A high EMA Score does not necessarily imply that an estimate is 'accurate' but that the information used allows us to understand and interpret the requirements with a degree of assurance. The value of an estimate should still be subjected to an appropriate level of 'challenge' including Verification and Validation in line with accepted good practice.

Note: Verification reviews that the calculations are appropriate and correct.
Validation reviews the appropriateness of the assumptions used.
(Validation and Verification are discussed further in Section 3.3.16)

However, one of the issues with the EMA definitions as they stand is that they might appear to be rather one dimensional, but in reality this is not true, as they require the estimator 'to read between the lines'. This opens up the question of the degree of subjectivity applied by one estimator in comparison with another. One potential way around this is to consider breaking down the subjectivity into a number of parts. If we read the original EMA Guideline Definitions we will see that they have three implied elements or dimensions to their Maturity:

1. Clarity of Task Definition
2. Task Familiarity
3. Quality of Data Accessed

	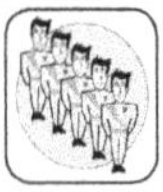	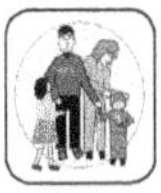	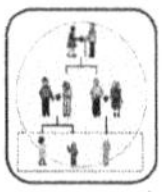	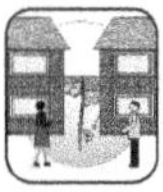	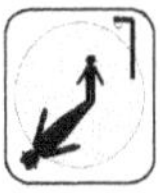	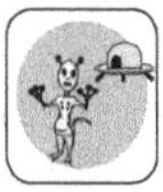
EMA	Clones	Blood Relatives	Close Cousins	Near Neighbours	Strangers	Aliens
9	✓	✓				
8		✓				
7		✓	✓			
6			✓	✓		
5			✓	✓		
4			✓	✓		
3				✓	✓	
2					✓	
1					✓	✓

Figure 3.2 Alignment of Expected EMA Levels with Different Product/Service Familiarity

From each of these three dimensions, we might define a Maturity Contribution (as shown in Table 3.3):

The EMA Score would then be the simple summation of the less subjective assessment by each dimension. This gives rise to 48 theoretical combinations, of which only some 37 are ever likely to be created, or be valid. Table 3.4 illustrates the 11 that are unlikely to occur.

Figure 3.3 shows the likely distribution of EMA Scores using this Dimensional Analysis Approach. Whilst the theoretical distribution is Normally Distributed (see Volume II Chapter 4), the Realistic Distribution is less centralised. The rare cases of scoring very high or very low scores are not unreasonable – without being unduly optimistic or pessimistic, it is rare that all the good things in life happen together, nor do all the bad things – we tend to get a mix in the middle. (*It just feels like that sometimes, doesn't it?*)

Caveat augur

Initially, estimators can get a little twitchy about EMAs …

Any Estimate Maturity Assessment should not be considered to be a measure of the integrity, professionalism or competence of the estimator who produced the estimate. It is a measure of the maturity of the information used at the time, or in the time available to produce the estimate; this is often outside the control of the estimator.

Table 3.3 Potential Dimensions of Estimate Maturity

Score	Clarity of Task Definition	Task Familiarity	Quality of Data Accessed
3	Precise agreed definition (no significant change expected)	Very familiar product or process (i.e. repeat or continuing business)	Repeatable, validated data or metrics
2	Good clear but unconfirmed definition (some risk of change)	Familiar type of product or process (e.g. new business)	Single point analogy, or unvalidated variable data or metrics
1	Outline definition only with detail yet to be confirmed or agreed (high likelihood of change)	Unfamiliar or infrequent product or process (e.g. emerging business)	Use of third party data of unknown provenance (unvalidated), or use of Expert Judgement
0	Vague description with ambiguous scope (high degree of change expected)	No experience of an innovative product, process or technology	N/A (any Estimate requires at least one data reference point)

Table 3.4 Estimate Maturity Assessment Dimensions Unlikely to Occur in Combination

EMA	Definition	Familiarity	Data Quality	Unlikely combination due to …
7	3	1	3	Data Quality inconsistent with Task Familiarity
6	3	0	3	Data Quality inconsistent with Task Familiarity
6	2	1	3	Data Quality inconsistent with Task Familiarity
6	0	3	3	Task Familiarity inconsistent with Task Definition
5	2	0	3	Data Quality inconsistent with Task Familiarity
5	1	1	3	Data Quality inconsistent with Task Familiarity
5	0	3	2	Task Familiarity inconsistent with Task Definition
4	1	0	3	Data Quality inconsistent with Task Familiarity
4	0	3	1	Task Familiarity inconsistent with Task Definition
4	0	1	3	Data Quality inconsistent with Task Familiarity
3	0	0	3	Data Quality inconsistent with Task Familiarity

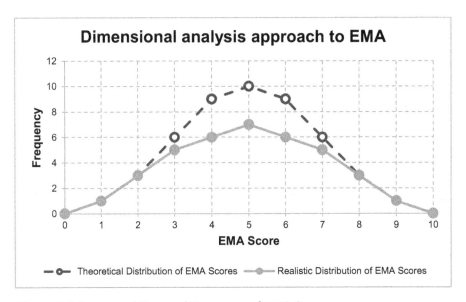

Figure 3.3 Range and Expected Frequency of EMA Scores

Unreasonable demands like '*I want an estimate and I want it now!*' do not lead to good robust Estimates, and may result in a lower EMA score than could have been achieved. However, there will be many situations when a low scoring EMA is all that could be reasonably expected – such as in estimates for concept development using innovative processes; by definition, any estimate will be immature based on the information available.

3.2.2 Schedule Maturity Assessment (SMA)

We can easily extend the principles under-pinning the EMA to assess the maturity of a schedule by adopting and adapting the defi-nitions of maturity from a scheduling rather than cost estimating perspective as shown in Table 3.5.

> **A word (or two) from the wise?**
>
> '*Time is money.*'
> Benjamin Franklin
> American Politician
> 1706–1790

There is a level of partial correlation (*we'll have to wait until Volume II Chapter 5 for that particular treat*) between cost and schedule epit-omised by the old adage of Benjamin Franklin.

These measures can be extended further to mirror the enhanced definitions shown previously for Estimate Maturity Assessments in Table 3.3.

Table 3.5 Schedule Maturity Assessment levels

Level	Schedule based on ...
SMA9	Precise definition with demonstrated task/activity durations and interconnectivity of the exact same nature to the required Schedule
SMA8	Precise definition with demonstrated task/activity durations and inter-connectivity for a well-defined similar task/activity duration to the required Schedule
SMA7	Precise definition with validated metrics for a similar task/activity duration and task dependencies to the required Schedule
SMA6	Good definition with metrics for a defined task/activity similar to that for the required Schedule including task/activity dependencies
SMA5	Good definition with historical information comparison for a defined task/activity, similar to that for the required Schedule, including task/activity dependencies
SMA4	Defined scope with good historical information comparison with the required Schedule including basic understanding of task/activity dependencies
SMA3	Defined scope with some appreciation of task/activity inter-connectivity, but with poor historical data comparison with the required Schedule
SMA2	Poorly defined scope and little understanding of task/activity inter-connectivity with poor historical data comparison with the required Schedule
SMA1	Poorly defined scope with no historical data comparison with the required Schedule, and little understanding of the inter-connectivity of tasks/activities

3.2.3 Cost and Schedule Integration Maturity Assessment (CASIMA)

The natural extension to having both an Estimate Maturity Assessment and a Schedule Maturity Assessment is to assess the combined maturity of cost and schedule planning. This allows us to gain a first impression of the maturity of the project's cost and schedule integration or CASIMA.

A mature cost estimate can be ruined by poor attention to the schedule on which it is based ... and vice versa. This frequently results in one or more of the following undesirable consequences:

- cost overruns
- excessive overtime
- excessive idle time
- schedule slippage

In order to measure the CASIMA Level, we can simply take the Geometric Mean of the EMA and SMA components. (*We will be covering what a Geometric Mean is in*

Table 3.6 Cost and Schedule Integration Maturity Assessment (CASIMA)

		Schedule Maturity Assessment								
		1	2	3	4	5	6	7	8	9
Estimate Maturity Assessment	1	1.0	1.4	1.7	2.0	2.2	2.4	2.6	2.8	3.0
	2	1.4	2.0	2.4	2.8	3.2	3.5	3.7	4.0	4.2
	3	1.7	2.4	3.0	3.5	3.9	4.2	4.6	4.9	5.2
	4	2.0	2.8	3.5	4.0	4.5	4.9	5.3	5.7	6.0
	5	2.2	3.2	3.9	4.5	5.0	5.5	5.9	6.3	6.7
	6	2.4	3.5	4.2	4.9	5.5	6.0	6.5	6.9	7.3
	7	2.6	3.7	4.6	5.3	5.9	6.5	7.0	7.5	7.9
	8	2.8	4.0	4.9	5.7	6.3	6.9	7.5	8.0	8.5
	9	3.0	4.2	5.2	6.0	6.7	7.3	7.9	8.5	9.0

Table 3.7 Potential Implications of CASIMA Scores

		Schedule Maturity Assessment								
		1	2	3	4	5	6	7	8	9
Estimate Maturity Assessment	1							Indicative of		
	2	Indicative of						Dysfunctional		
	3	Shared Reality						Estimating		
	4									
	5					Indicative of				
	6					Co-ordinated				
	7	Indicative				Cost & Schedule				
	8	of Dysfunctional				Planning Activity				
	9	Schedule Planning								

Volume II Chapter 2; I know, I can hardly wait myself, but if you really can't wait, in this particular case, multiply the two elements together and take the square root of the product.) By using the Geometric Mean rather than a simple average we will avoid getting an artificially high CASIMA, where either the EMA or the SMA is high in relation to the other. Table 3.6 summarises the results and rounds the intersecting values to the nearest integer.

If we reflect on what Table 3.6 is inferring, then we might come up with conclusions similar to those in Table 3.7.

3.3 Good Practice Spreadsheet Modelling (GPSM)

In many organisations the toolset of choice for estimating is a spreadsheet. In this book we have used Microsoft Excel as the means of demonstrating many of the numerical techniques discussed. The reason for this is that its strengths are its convenience, ease of use and flexibility; it is probably the low cost option too. However, its weakness is also its ease of use and flexibility, leading in many cases to spreadsheets that are badly designed, difficult for others to follow and modify and prone to errors. Many of the negative consequences can be mitigated without losing the positive benefits of those same attributes ... so long as we follow some simple 'Good Practice' Guidelines recommended by organisations such as the European Spreadsheet Risk Interest Group (EuSpRIG).

We should recognise that no 'one size fits all', but instead we should look to each of our organisations and encourage them to write or adopt a procedure that encompasses the good practice principles for Spreadsheets in line with the six stages of modelling recommended by Read and Batson (1999):

1. Scope
2. Specification
3. Design
4. Build
5. Inspection
6. Use

How many of us, or our colleagues, go straight into 'Build' and then 'Use', and pay little attention (*if any, being cynical about it*) to the other four? (*Yes, quite a show of hands there.*)

There are some key principles to which we should all try to adhere, and these are in line with or support our principles of TRACEability. These Good Practice Principles can be summarised as MUST (*it would have been better had these spelt out SHOULD, but MUST will have to do*):

- **Maintainability** ... which supports estimate or model Inheritability and implies Simplicity

- **Usability** … which infers that it is easy to use (Simplicity again) and well-documented
- **Security** … which guards against accidental corruption of the spreadsheet without denying Accessibility
- **Transparency** … which support estimators' need for Auditability (Verifiability & Validatability)

These will be used in the next section to develop an IRiS Score (Inherent Risk in Spreadsheets). Let's look a little closer at some of the good practice guidelines.

3.3.1 Level of documentation (T, M)

Based on the first five stages and inputting into the sixth stage, there should be documentation to support the spreadsheet. In many instances this can be included in the Excel Workbook itself, especially for smaller models, as one or more context tabs or worksheets. These should include (but not limited to) the following:

- Statement of Purpose
- Scope of what is included and excluded
- Any limitations on its use
- User Instructions
- Maintenance Guide (for really big models)
- Version Control for the Model Structure
- Version Control for the Data
- Index and purpose of each Worksheet
- Model Map
- Summary of Key Inputs and Outputs
- Index of Named Ranges

3.3.2 No hidden worksheets, columns or rows (T, M)

There should be no hidden worksheets, columns or rows in our Excel Workbooks. Doing so fails the principle of Transparency and Maintainability. What are we trying to hide or protect?

If we are doing it to stop someone accidently deleting key data inputs or calculations then there are better ways we can do this using Cell and Tab Colours as signals (see Section 3.3.3), or by taking some simple Security precautions (see Section 3.3.4).

What's the risk of hiding columns or rows? Someone may inadvertently delete the column or row, losing data or functionality in the process.

If the row or column is merely tidying up blank real estate in the spreadsheet, then we might consider minimising the column width or row height to something small (but greater than zero) instead. Visually, it highlights that there is a row or column but that it is not significant.

Why have we not just said delete blank rows or columns? We'll discuss that in Section 3.3.13.

3.3.3 Colour coded cells and worksheet tabs (U, S)

How many times have we opened up someone else's spreadsheet and thought, '*Hmm, I wonder what all these cells do, and which I can change, and where are the final outputs?*' Wouldn't it be helpful if all the input cells were grouped together, and all the calculation cells etc., either on separate Worksheets or in separate areas of a single Worksheet. Table 3.8 illustrates the concept. However, this does not mean that we cannot have checksum values on Input Sheets, nor some input parameters clearly highlighted on a calculation worksheet.

Table 3.8 Potential Worksheet Tab Colour Coding

Worksheet Type	Possible Tab Colour	Explanatory Notes
Contextual or Administrative Information	Default Grey	These worksheets are used to summarise the working assumptions, version control information, Colour Legend etc. It is often appropriate to have separate worksheets for different contextual and administrative functions
Inputs	Pale Yellow	These are worksheets that contain predominately input data and constants. There could be more than one of these, depending on model complexity, to segregate data appropriately (e.g. Technical Inputs, Manufacturing Inputs etc.)
Sensitivities	Pale Orange	These are worksheets that contain input data sensitivity ranges. These allow users to flex the values of inputs to test the sensitivity of change on the results
Calculations	Pale Green	These are worksheets that contain predominately calculation data. There could be more than one of these, depending on model complexity, to segregate functionality appropriately (e.g. Technical Calculations, Manufacturing Calculations etc.). This could include data cells generated by Array Formulae and/or Macros where used
Output Data	White	These are cells which refer to formatted output reports which draw information from various input and calculation worksheets. It is recommended that there is a separate worksheet for each distinct report type

Worksheet Type	Possible Tab Colour	Explanatory Notes
Random Numbers	Pink	These worksheets hold data generated at random, such as output from Monte Carlo Simulation (MCS). These may or may not be generated automatically by the MCS Application as an add-in to Microsoft Excel
Dynamic Links to External Data	Turquoise	Where large quantities of Dynamic Links to External Data are included, it is recommended that these are created on their own worksheets so that separate data validation can be enabled more easily
Validation Summary	Black	It is considered to be good practice to summarise all validation checks on a single worksheet (a one-stop-shop for error checking)
Mixed Cell Types	Default Grey	Where mixed cell types are included on a single worksheet, a neutral colour should be chosen; default grey would be appropriate

Table 3.9 Potential Cell Colour Coding

Cell Type	Example Cell Colour Format	Explanatory Notes
Input	Pale Yellow with Black Font	These are cells which a user may change, representing assumption variables. There may be limitation on the values input but these can be controlled using Data Validation (Section 3.3.15)
Sensitivities	Pale Orange with Black Font	These are specific inputs that enable users to flex other inputs to test the sensitivity of results. Examples might include Minimum and Maximum values
Constant	Blue with White Font	These are special types of inputs which the user is not expected to change. They may relate to published data such as government furnished historical escalation rates, or inviolate constants such as the number of months in a year
Calculation	Pale Green with Black Font	These are cells which contain formulae or use Microsoft Excel's built-in functions, or simply just refer to the value of another cell e.g. =A7
Output Data	White with Black Font	These are cells which refer to previous inputs of calculations cells and depict formatted output reports
Array Calculation	Mid Green with White Font	These are special calculation cells which use Array Calculations (see Section 3.3.8)

(Continued)

Table 3.9 *(Continued)*

Cell Type	Example Cell Colour Format	Explanatory Notes
Random Numbers	Pink with Black Font	These are cells that hold data generated at random, such as output from Monte Carlo Simulation (MCS). These may or may not be generated automatically by the MCS Application as an add-in to Microsoft Excel.
Macro Generated Data	Lavender with Black Font	It is advised that Macros are only used to generate data in specific circumstances in Microsoft Excel (see Section 3.3.7). If Macros are used, then it is recommended that their output be coloured distinctly to avoid being accidentally over-written. Note: Certain internal Excel facilities such as Data Analysis and Solver will generate data using internal macros
Dynamic Links to External Data	Turquoise with Black Font	Dynamic Links to External Data are discouraged (see Section 3.3.9). Where they are used, they should be clearly highlighted
Column Headers	Grey with Black Font	Column and Row Headers and Labels should be identified with an appropriate colour. This might include headers or labels generated by reading
Row Labels	Grey with Black Font	or combining other data. As part of the internal standard, it may be advantageous to differentiate those input directly to those that are generated by different shades of grey
Notes	White with Black Font	Any notes added by the user or spreadsheet developer can be added
Named Range Headers	Red with White Font	To avoid accidental corruption of Named Ranges (see Section 3.3.10), it is recommended that such ranges are clearly labelled
Data Validation Pass Signal	Bright Green with Black Font	It is considered good practice to include error validation checks within a spreadsheet. Where these error checks pass or fail, they should be flagged accordingly (using Conditional Formatting, for example)
Data Validation Fail Signal	Bright Red with Black Font	
Data Validation Warning Signal	Amber with Black Font	Conditional Warnings can be added if a value is close to or exceeds a given hurdle level, but is not incorrect as a calculation

Perhaps even more useful is the idea that all cell types of the same type should be coloured the same, i.e. all the inputs were one colour, all the calculation cells were another colour. To make it even more useful, wouldn't it be a good idea if every organisation had a standard colour template for spreadsheet cells? Table 3.9 illustrates the concept (the colours are down to organisational preference and/or governance).

3.3.4 Locked calculation cells and protected worksheets and workbooks (S)

To avoid accidental corruption of the workbooks, the following safety precautions are recommended:

- All Cells other than User Input Cells (pale yellow in Table 3.9) should be locked in Microsoft Excel. Microsoft encourage this by setting 'locked' as the default condition for any cell once Worksheet Protection is switched on. In reality, therefore, we have to perform a conscious act to unlock input cells. (This can be done with Stylesheets, or manually using the menu structure.)
- We should then enable Worksheet Protection to enable the Cell locking feature. This will allow users to access the unlocked input cells only. We will have had the option to prevent or allow users to access any cell to allow them 'read only' permissions to understand any functions or logic used, but not to edit them. This is good practice from a Transparency perspective. There is also an option to apply a Password when protecting the worksheet but this does introduce the risk that someone will forget the password and lock everyone out from maintaining the model in the future – Single Point of Failure! (We might want to consider Worksheet Protection without Passwords.)
- Similarly, we should consider applying Workbook Protection (without a Password) to prevent someone accidentally deleting a Worksheet, adding rows or columns, and corrupting the model. (*Can we honestly say we've never done it ourselves?*)

It can be argued that there is no point in applying Workbook and Worksheet Protection without a Password because someone can always turn it off. This is true, but then it will have been a deliberate act ... sabotage even, if they had no legitimate right to do so.

3.3.5 No hard-coded constants unless axiomatic (M)

It is considered to be bad practice to include constants within a calculation formula. This violates the Maintainability principle and may result in future errors if the constant needs to be changed but is 'missed'. It is better practice to include the constant in a separate cell and to refer to it in a calculation.

However, where the 'constant' is an axiomatic value in a calculation that will never change, then we can probably relax that guideline. For instance, the area of a square will always be the square

> An Axiom is a statement or proposition that is generally accepted as being self-evidently true at all times, requiring no further proof.

of one of its sides. In some instances, in the interests of Transparency, we may still be justified in defining a constant in a separate Constant Cell (blue background with white text in Table 3.9) to improve readability, even though the value will never change. For instance, even though there will always be 12 months in a year, dividing by a cell labelled or even Named 'MonthsInYear' explains why we are dividing an annual value by 12.

On the left hand side of Table 3.10 we show a calculation which includes two hard-coded constants. One of these represents the Average Escalation Factor per Year of 1.031 (or 3.1% per annum), and the other is the Baseline Year used to calculate how many years of escalation we need to include in order to estimate our outturn costs.

On the right hand side we have extracted these two values and treated Baseline Year as a true invariable quantity depicting it with a Blue Background and White Text (as per Table 3.9, *albeit only in black and white here!*). In terms of the Average Escalation Factor per year, we have treated this as a regular input variable … even if we have no intention at present of changing it.

In Volume III Chapter 2 we will make reference to the Generalised Solution for a Quadratic Equation. In cases like this the constants and powers will never change so it is acceptable to hard-code them in the calculation.

Another case where we can relax this principle is where there are parameter values in Excel functions that require a constant. For example, the function **MATCH(*lookupvalue, lookup_array, match_type*)** requires the *match_type* parameter to be -1, 0 or 1; no other value will work. We could set up three Named Ranges for these called NextGreaterMatch, ExactMatch, NextSmallerMatch or something similar that we can

Table 3.10 Moving Constants into Addressable Cells

	A	B	C	D	E
1					
2					
3	Baseline Unit Cost		£ 2.734 k		
4	Delivery Year	Units per Year	Baseline Cost	Cost @ Outturn	
5	2013	5	£ 13.670 k	=C5*1.031^(A5-2010)	
6	2014	10	£ 27.340 k	£ 30.891 k	
7	2015	12	£ 32.808 k	£ 38.218 k	
8	2016	12	£ 32.808 k	£ 39.403 k	

✘ Hard-coded constants in the calculation

	A	B	C	D	E
1	Ave Escalation Factor		1.031	per year	
2	Baseline Year		2010		
3	Baseline Unit Cost		£ 2.734 k		
4	Delivery Year	Units per Year	Baseline Cost	Cost @ Outturn	
5	2013	5	£ 13.670 k	=C5*C$1^(A5-C$2)	
6	2014	10	£ 27.340 k	£ 30.891 k	
7	2015	12	£ 32.808 k	£ 38.218 k	
8	2016	12	£ 32.808 k	£ 39.403 k	

✔ Putting the constants in addressable cells

For the Formula-philes: General Solution of a Quadratic Equation

Consider a general Quadratic Equation of the form:	$ax^2 + bx + c = 0$
The solution to which can be found by:	$x = \dfrac{-b \pm \sqrt{b^2 - 4ac}}{2a}$

---

use throughout the workbook (*or is that becoming a little too fastidious?*) We will revisit this discussion in Section 3.3.10 on Named Ranges.

3.3.6 Left to Right and Top to Bottom readability flow (U)

In Western Cultures it is normal practice to read from Left to Right and from Top to Bottom. The basic logic of a spreadsheet should also flow in this way. However, enforcing this as an absolute is often unnecessarily bureaucratic, and common sense should be allowed to prevail. For instance, we would expect to summarise 12 columns of monthly data on the right as a total for the year. We may then want to calculate the percentage value for each month relative to this annual total. This would require us to read from the write, breaking the Left to Right flow.

We can use the inbuilt capability within Microsoft Excel to check whether we are following this or not. If we open the Formula Auditing Tools and click on the Trace Precedents and/or Trace Dependents for any cell we will get blue arrows that show the direction of flow of information in the spreadsheet calculations.

In our example in Table 3.11, the data flow fails to comply with the principle of Left to Right, Top to Bottom Readability. For instance, Cell F8 is required by Calculation Cell E2; the flow is up and to the left. The same is true for the whole of Column F. The other flows all adhere to Left to Right and Top to Bottom.

We can re-arrange this data in a number of ways to comply with Left to Right, Top to Bottom readability principle as illustrated in Tables 3.12 and 3.13. The first of these keep related sets together (Standard Time Inputs and Calculations, Learning Curve Inputs and Calculations) before combining them to arrive at a total Estimate for the First Article Hours. The second of these keep all the inputs together on the left and calculations on the right.

In those cultures which do not follow a Left to Right Reading/Writing convention, the equivalent alternative convention might be considered.

Table 3.11 Data Flow Fails to Comply with Principle of Left to Right and Top to Bottom Readability

	A	B	C	D	E	F
1	Work Centre	Parts Count	Average Standard Time (Hours) per Part	Estimated Total Standard Time (Hours)	First Article Hours	
2	Total			69.79	169.07	
3	Machining	15	2.75	41.25	61.05	
4	Fabrication	24	0.67	16.08	35.20	
5	Electrics	4	0.75	3.00	10.64	
6	Assembly	43	0.22	9.46	62.18	
7						
8	Work Centre	Learning Rate	Learning Exponent	Baseline Unit	Current Average % Performance to Standard @ Baseline	First Article Factor
9	Machining	95%	-0.0740	100	95%	1.4801
10	Fabrication	90%	-0.1520	100	92%	2.1889
11	Electrics	85%	-0.2345	100	83%	3.5469
12	Assembly	80%	-0.3219	100	67%	6.5733

Table 3.12 Data Flow Complies Fully with Principle of Left to Right and Top to Bottom Readability (1)

	A	B	C	D	E	F	G	H	I	J
1	Work Centre	Parts Count	Average Standard Time (Hours) per Part	Estimated Total Standard Time (Hours)	Learning Rate	Baseline Unit	Learning Exponent	Current Average % Performance to Standard @ Baseline	First Article Factor	First Article Hours
2	Machining	15	2.75	41.25	95%	100	-0.0740	95%	1.4801	61.05
3	Fabrication	24	0.67	16.08	90%	100	-0.1520	92%	2.1889	35.20
4	Electrics	4	0.75	3.00	85%	100	-0.2345	83%	3.5469	10.64
5	Assembly	43	0.22	9.46	80%	100	-0.3219	67%	6.5733	62.18
6	Total			69.79						169.07

Table 3.13 Data Flow Complies Fully with Principle of Left to Right and Top to Bottom Readability (2)

	A	B	C	D	E	F	G	H	I	J
1	Work Centre	Learning Rate	Baseline Unit	Current Average % Performance to Standard @ Baseline	Parts Count	Average Standard Time (Hours) per Part	Estimated Total Standard Time (Hours)	Learning Exponent	First Article Factor	First Article Hours
2	Machining	95%	100	95%	15	2.75	41.25	-0.0740	1.4801	61.05
3	Fabrication	90%	100	92%	24	0.67	16.08	-0.1520	2.1889	35.20
4	Electrics	85%	100	83%	4	0.75	3.00	-0.2345	3.5469	10.64
5	Assembly	80%	100	67%	43	0.22	9.46	-0.3219	6.5733	62.18
6	Total						69.79			169.07

3.3.7 Avoid data generated by macros ... Unless there is a genuine benefit (S, T)

Macros that use absolute Cell, Column or Row references are at high risk of corruption if a worksheet column or row is added or cells are relocated by dragging and dropping. Hence, it is recommended that as normal practice, macros are used for navigation only unless there is a real tangible benefit to using a Macro, such as for the processing of large quantities of data for analysis which do not update automatically when data is updated; an example in Microsoft Excel might be Linear Regression using the Data Analysis Add-in.

Where macros are used, cells and ranges should be 'Named' (see Section 3.3.10) in place of absolute references. This will ensure the integrity of the Macro should data be moved or columns or rows added.

Note: Where we have used internal features of Excel that use Macros 'behind the scenes' such as Data Analysis and Solver, we can safely assume that these principles have been applied. However, in some cases these internal Macros generate calculation formulae, but in others they just 'paste' a value. By colouring all Macro-generated data in a unique colour, we are signalling that there may be data that will not refresh automatically if constituent input cells are changed. This should also be used for data generated by special functions and features within Microsoft Excel, such as output generated by the Data Analysis Add-in.

Suppose we have a Macro that contains code to apply the appropriate Charging Rate to the Estimated Machine Hours and puts the net cost in Cell AF3:

```
Range("AF3").Select
    ActiveCell.FormulaR1C1 = "=RC[-1]*R[7]C"
```

The last of these lines looks at the location of the Hours and Rates data relative to Cell AF3; in this case one column to the left on the same row for the Hours, and seven rows below in the same column for the Rate.

If we insert an earlier row or column this code will fail. Instead we should give the three cells Names (see Section 3.3.10); we can then replace this with the more robust code:

```
Range("MachineCost").Select
    ActiveCell.FormulaR1C1 = "=MachineHours*MachineRate"
```

3.3.8 Avoid Array Formulae (T, U, M)

These are those very clever formulae that we may sometimes see in squiggly brackets ... {}.

The squiggly brackets or 'braces' are not typed in by the user but are created by Excel when the user hits 'ctrl + shift + enter' simultaneously with the appropriate functions. If we then edit the Array Formula and incorrectly press 'enter' on its own, we are likely to create an error. If we are lucky, we will get a clear #VALUE error. If we are not so lucky, we may just get an incorrect value generated.

Array Formulae do not support being 'dragged and dropped' very well either, and can produce fairly random looking answers!

Often there is a simpler way in Excel to perform a calculation than to use an Array Formula. For example:

The Array Formula: **{=SUM(A2:A10*B2:B10)}**
can be substituted by the simpler and somewhat eponymous **SUMPRODUCT**
function **=SUMPRODUCT(A2:A10,B2:B10)**

The Array Formula: **{=SUM(IF(YEAR($C5:$H5)=C$14,$C7:$H7))}**
can be substituted by the simpler and more intuitive **SUMIF** function if we
include the Year in Row 4 extracted from the date in Row 5 i.e. **C4=YEAR(C5)**
etc. **=SUMIF($C$4:$H$4,C$14,$C7:$H7)**

There will be occasions (but probably quite rare) where Array Formulae are the only practical way of achieving our ends. An example we use in Volume III Chapter 4 are Prediction and Confidence Limits around a Multilinear Regression. In these few cases we can use Array Formulae so long as we document what they do, and colour the cells a different shade of the colour chosen for normal calculations. In Table 3.8, we used the example of pale green for normal calculations; if we do use Array Formulae, we might colour these cells a brighter or deeper green.

We should use Array Formulae with caution, and only where we have a genuine need, not just because we can! The next user may not be aware of what they are and how they are used; recall the Spreadsheet Inheritability Principle, which is a part of Maintainability.

3.3.9 Avoid dynamic links to external data (S)

Microsoft Excel allows us to compartmentalise information and calculations into different Workbook files and to link them together rather than recreate all the functionality in a single huge file, or to link to files generated by other people. This can be extremely useful … but also dangerous for the integrity of our files.

Suppose we were to freeze our assumptions that we are using for our estimate, but someone changes an assumption value in a 'feeder' spreadsheet. There is a real risk that this change will flow through unchecked into our spreadsheet changing our estimate in the process, and we probably should not want that.

It is safer practice to extract data as values rather than links from external sources (*Cut, Paste Special, Values*). However, we may want to know for a second iteration of an estimate when someone has updated their contribution. In this case we may want to consider a link to that data as a Checksum value (coloured Turquoise as suggested in Table 3.8, or whatever colour we have chosen) with an ErrorVet against the original 'Cut and Pasted' value rather than as an active dynamic link. To put it another way, using a passive link instead mitigates the risk to our spreadsheet from someone else's action.

3.3.10 Use Named Ranges for frequently used table arrays (M, U)

To aid readability of complex calculations, and to protect against changes in row and column data positions, we can use another Microsoft Excel feature of assigning a Name to a range or block of data, or even a single cell. This is particularly helpful if we are referring to particular sets of data or calculations frequently, giving improved integrity as well as improved readability.

To safeguard these Named Ranges, it is a good idea to use a Named Range Header (as suggested in Table 3.14) as a signal to others not to insert rows or columns through the Named Range. For larger Named Ranges, it is also a good idea to include a column and/or row counter as illustrated in the example in Table 3.14 to improve transparency when using **VLOOKUP, HLOOKUP, INDEX** and other similar functions. These counters lie outside the Named Range.

VLOOKUP(Electrics, ChargingRates, 4, FALSE) will return the value £67.88 / Hr. The astute amongst us will have spotted that we have included a hard-coded constant here for the column number parameter. In many cases, this will seem like the practical thing to do, but it does introduce a risk if someone inserts a column for 2015. We can often mitigate risks like these by replacing the hard-coded parameter, 4, with a function that looks up its current position in an array such as **MATCH(2018,ChargingRate-Year,0)**. Here, we will have first had to define the Named Range 'ChargingRateYear' to be the first row of the 'ChargingRates' Named Range. The zero parameter in the **MATCH** function signals that we want an exact match.

Table 3.14 Example Named Range

ChargingRates				
1	2	3	4	5
Rate Pool	2016	2017	2018	2019
Machining	£ 89.32 / Hr	£ 90.97 / Hr	£ 93.43 / Hr	£ 95.29 / Hr
Fabrication	£ 61.13 / Hr	£ 63.76 / Hr	£ 64.37 / Hr	£ 65.97 / Hr
Electrics	£ 65.21 / Hr	£ 67.23 / Hr	£ 67.88 / Hr	£ 70.52 / Hr
Assembly	£ 73.89 / Hr	£ 78.77 / Hr	£ 80.23 / Hr	£ 81.75 / Hr

3.3.11 Use full syntax within Excel (M)

With many of its functions Microsoft Excel allows some parameters to be optional. Where it does this, it assumes a default value. It may be that default value is not the one we would intuitively expect … and we might then get surprising results!

In Sections 3.3.5 and 3.3.10 we looked briefly at the Microsoft Excel **MATCH** function, the last parameter of which match_type is optional; where we don't specify it, Excel will assume a default of one, i.e. the largest value that is less than or equal to the value we want to look up. We may have been expecting that an Exact Match would be the default; if it cannot find an Exact Match then it would return an error, but by assuming a default of one it reduces the risk of an error occurring. (If the lookup_value is less than the first value in the list then it will still return an error.)

Avoid the unintended … always use the full Excel Syntax.

3.3.12 Break complex calculations into smaller simpler steps (T, M)

Complex calculations are those that use advanced functions within Microsoft Excel (such as Array Formulae), or multiple condition statements (**IF, AND, OR** functions). These increase the chance of errors being built into the spreadsheet, and reduce transparency.

Someone once told me that a formula in Excel should be no longer than my thumb. (We could call this the '"Rule of Thumb" Rule of Thumb'. Well, if we stand far enough back from our screens, they may appear to be no longer than our thumbs, but they will also get even less readable from a distance!

The sensible approach is to break long calculations down into smaller, more manageable steps. This improves integrity and transparency.

Also, we will all probably know that we can write a calculation in several different ways in Excel, all of which are correct and give the same answer. It is recommended that we use the same 'coding' for blocks of calculations with the same functionality in which only the row and column references change:

e.g. If the rows are different functions, and the columns are different years, the calculation block of cells applying charging rates to different functions across a number of years will be the same, other than the specific references to rows and columns. Don't change the calculation formula unnecessarily.

3.3.13 Column and row alignment across worksheets (T)

To aid readability and spreadsheet development it is recommended that comparable columns and rows of data are aligned within and across worksheets. For instance, suppose we have a worksheet with the resource hours estimated for a different Work

Breakdown Element in each row, and the Months across the columns. In another worksheet we have the corresponding data for the costed hours. It improves readability if the same rows and columns are used in the two worksheets for the corresponding data pairs.

It makes the task of Validating and Verifying much easier.

3.3.14 Unambiguous units of measure (U)

Each cell value should have an associated unit of measure. This can be done in one of several ways depending on circumstances and personal preference:

- As a Column Header
- As a Row Label
- As row values in a separate column (suitable for mixed data types in a single column)
- As an integral part of the cell formatting (like currency and percentage signs)

We cannot assume that just because we know what a number in a cell meant when we created a spreadsheet that another user will be able to read our minds. A common error made of this nature that we should guard against is that people sometimes forget to tell us that the cost or hours etc. are in thousands, millions or billions! A simple £k, $m or €b in the column header or cell formatting will resolve that.

3.3.15 Input data validation (U)

Ever had that problem of people mistyping data into input columns, or inputting values that fall outside an acceptable range? We can use Microsoft Excel's Data Validation to control these so that we don't get errors or invalid results.

For example, suppose our model allows another user to input a functional group for which they want to calculate the relevant costs using the appropriate charging rate. It may be important that they use the same word or spelling that we use otherwise the model will fail. Wouldn't it be a good idea to prevent them inputting 'Manufacturing' or 'Operations' or 'Prod' if the model requires them to input 'Production'?

Similarly, if we expect integer values in the range of 1 to 7 (for example), we can … and should … prevent them inputting 13.84, or whatever.

3.3.16 Independent model verification and validation (S)

Ever had that problem of seeing something that you expect to see rather than what is actually there? (*No? Perhaps it's just me.*) It is very difficult to review your own work effectively for that reason; even though I had proofread every chapter for errors in this volume (and the others), other people will have still found mistakes that I have missed.

The same must be said of spreadsheets. It is considered to be essential good practice that someone independently verifies and validates (V&V) the spreadsheet; ideally someone not connected with the current project. It doesn't have to be the same person who both verifies and validates the spreadsheet. It is not unreasonable for us to ask what the difference is, after all they are often used interchangeably and have different nuances depending of their contest. In this context we mean the following:

Definition 3.2 Spreadsheet or Model Verification

Verification is the process by which the calculations and logic of a spreadsheet or model are checked for accuracy and appropriateness for their intended purpose.

Definition 3.3 Spreadsheet or Model Validation

Validation is the process by which the assumptions and data used in a spreadsheet or model are checked for accuracy and appropriateness for their intended purpose.

If we have a robust estimating process that routinely challenges or validates the assumptions we are using, the validation of a spreadsheet boils down to whether someone has included all the relevant data assumptions, and have they input them correctly?

In terms of independence and degree of verification performed, this can range:

• From the person at the next desk just doing a 'quick peer review' of the key information flows, and checking some calculations at random.
• To someone external to the project who has been trained formally in the arts and craft of V&V assessment, and checks every single calculation and input value. (*What, no volunteers?*) Despair not, there are Add-in tools for Microsoft Excel that will

enable this to be done more easily, diminishing the task to a verification of unique calculations only.

3.4 Inherent Risk in Spreadsheets (IRiS)

… or how to keep an eye of how robust our cost model is!

We can use the following as a self-assessment tool, or it can be used by someone performing a Validation and Verification of someone else's spreadsheet. It links directly with Good Practice Spreadsheet Modelling.

We can use the Good Practice Principles to develop a view of the Inherent Risk there is in our spreadsheet, by scoring how we have performed against each principle, and then taking an average:

- Follows Good Practice principles well – score 0
- Occasionally fails to follow Good Practice principles – score 1
- Several instances where Good Practice is not observed – score 2
- Multiple instances of failing Good Practice principles – score 4

Based on the descriptions in Table 3.15, we can view it on a Radar Chart (Figure 3.4). Here, the average score is 0.8, but there is an elevated risk associated with the principles of Maintainability and Security.

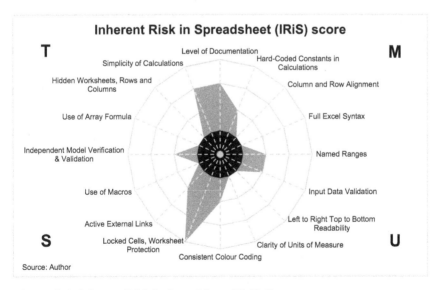

Figure 3.4 Inherent Risk in Spreadsheet (IRiS) Score

Table 3.15 IRiS Scores by Good Practice Spreadsheet Principles

Good Practice Feature	Risk Score				
	0 – Low Risk	1 – Medium Risk	2 – High Risk	3 – Very High Risk	
Level of Documentation	Full documentation available	Most documentation Available	Little documentation available	No documentation available	
Hard-Coded Constants in Calculations	No hard-coded constants in calculation cells except axiomatic formula values	Some instances of hard-coded constants in calculation cells of small spreadsheet	Some instances of hard-coded constants in calculation cells of large spreadsheet	Multiple instances of hard-coded constants in calculation cells	
Column and Row Alignment	All corresponding columns and rows align across multiple worksheets	Some corresponding rows do not align across multiple columns of a single worksheet	Some instances where corresponding columns and rows are not aligned across multiple worksheets	Several instances where corresponding columns and rows are not aligned across multiple worksheets	
Full Excel Syntax	Full syntax is being used for all Excel functions	Occasional instances of syntax being abbreviated for a small number of Excel functions	Multiple instances of syntax consistently being abbreviated for a range of Excel Functions	Multiple but inconsistent instances of syntax being abbreviated for a range of Excel Functions	
Named Ranges	Extensive use of Named Ranges, proportionate to spreadsheet size and complexity	Some use of Named Ranges, proportionate to spreadsheet size and complexity	Small simple spreadsheet with no Named Ranges	Large complex spreadsheet with no Named Ranges	
Input Data Validation	Extensive use made of Input Data Validation with User Help explanations	Wide use made of Input Data Validation with User Help explanations	Limited use of Input Data Validation	No use of Input Data Validation on a complex spreadsheet with multiple input variables	
Left to Right Top to Bottom Readability	Consistent flow of workbook logic from left to right, top to bottom and front to back	Generally good adherence to workbook logic flowing from left to right, top to bottom and front to back	Logic flow is inconsistent with a number of instances of jumping from right to left, or bottom to top or back to front	Logic flow is haphazard jumping frequently from right to left, or bottom to top or back to front	

Clarity of Units of Measure	All cells, columns, rows and worksheet tabs are clearly labelled with their units of measurement/scales	Most cells, columns, rows and worksheet tabs are labelled with their units of measurement/scales	Basic cell value measurement units are visible but scale units (k, m, b) are frequently missing	Cell value measurement unit scales are largely missing
Consistent Colour Coding	Organisation-approved standard colours are used consistently and correctly	Generally, cell colours are not used, or there are some inconsistencies in the application of standard colours	Non-standard colours are used instead of organisation-approved ones	Similar cell types are coloured differently
Locked Cells, Worksheet Protection	Input cells are unlocked, other cells are locked, worksheets and workbooks are protected but without password	Input cells are unlocked, other cells are locked, worksheets are password protected or the workbook is unprotected	Some worksheets or the workbook are unprotected, leaving those cells unlocked and vulnerable to change	All worksheets and workbook are unprotected, leaving all cells unlocked and vulnerable to change or deletion
Active External Links	There are no active links to external data sources (other than checksum values)	There are some active links that simply read external data sources	There are some active links that perform calculations on external data sources	There are multiple links to or high value links calculated from external data sources
Use of Macros	Any macros are used for navigation only	Macros are used to calculate cell values, but they all use Named Ranges rather than absolute cell references	Some macros are used to generate cell values, and logic uses absolute cell references rather than Named Ranges	Multiple macros are used to generate cell values, and logic uses absolute cell references rather than Named Ranges
Independent Model Verification & Validation	Independently V&V'ed by trained Assessor	Simple Peer Review (Dip Check)	Multiple Checksum Vets (Self-inspection only)	No checking performed
Use of Array Formula	No Array Formulae are used	One Array Formula is being used but explanatory text is available	One Array Formula is being used without documented notes	Multiple instances of Array Formulae being used
Hidden Worksheets, Rows and Columns	No hidden worksheets, columns or rows	Worksheet hidden but no hidden rows or columns	Some rows and columns are hidden	Multiple instances of hidden rows and columns
Simplicity of Calculations	All calculations are simple	Calculations are complex but documented	Multiple nested condition formulae are being used without documentation	Multiple nested condition formulae are being used without documentation

The bigger the IRiS Score, the more encircled the black pupil becomes, signifying a greater Inherent Risk in the Spreadsheet. The black pupil alone denotes Low Risk. We must recognise that a higher IRiS Score doesn't mean necessarily that the Spreadsheet has any errors at present, but there is a greater chance that errors could be introduced during its regular use, or if the spreadsheet is being changed to add or modify functionality.

The more that we can see of the IRiS, the more that we need to keep an eye on our spreadsheet for emergent errors.

3.5 Chapter review

At the end of the day (*or in this case chapter*) there are two important things that we must deliver as estimators:

1.   The value(s) we have estimated
2.   How we arrived at the estimate(s), otherwise known as the Basis of Estimate

These two cannot be separated; without a basis of estimate, the estimate might just as well be some random number floating around without a context; that is its destiny.

Transparency, Repeatability, Appropriateness, Credibility and being Experientially-based (TRACEable) are critical characteristics that define a good Estimate and Basis of Estimate.

In terms of assessing the robustness of the Basis of Estimate, we might like to consider adding an Estimate Maturity Assessment or EMA to our Estimate, or even a Schedule Maturity Assessment (SMA). This gives us an opportunity to highlight how comfortable we are with the level of information used, or the time allowed to us, in developing the estimate. However, it is important to realise that an EMA or SMA is not a reflection of the professionalism or integrity of the estimator or planner. If we combine these together we may find that a CASIMA or Cost and Schedule Integration Maturity Assessment points us at areas where the organisation is, or is not, working together effectively in producing an integrated view of cost and schedules. If we find ourselves in that invidious school report position of '*could do better*', then we may need to look again at the effectiveness of our processes and procedures.

Often the tool of choice, or sometimes just the tool to hand for many estimators, is a Spreadsheet application like Microsoft Excel (*other spreadsheet applications are available!*). Their strength lies in their flexibility, but therein also lies their greatest weakness. The Good Practice Principles for Spreadsheet Modelling have been promulgated for some time, yet collectively we still have a tendency to wade in and do our own thing! We can use the IRiS Scorecard and Graphic to raise awareness of the Inherent Risk in Spreadsheets we have built, and where those risks reside in terms of Maintainability, Usability, Security and Transparency (MUST). The more of the IRiS we can see, the more we need to keep an eye on the inherent risk in the spreadsheet.

References

Read, N & Batson, J (1999) *Spreadsheet Modelling Best Practice*, London, Institute of Chartered Accountants in England and Wales.

Smith, E (2013) 'Estimate Maturity Assessments', Association of Cost Engineers Conference, BAE Systems, London.

4 | Primary and Secondary Drivers; Accuracy and precision

4.1 Thank goodness for Juran and Pareto

> **A word (or two) from the wise?**
>
> The Pareto Principle is what separates ' ... *the vital few from the trivial many.'*
>
> **Joseph Juran**
> Management Consultant

In the 1940s Joseph Juran, Management Consultant and Quality guru, coined the phrase the *Pareto Principle* after the observations of Italian Economist Vilfredo Pareto (1906). Pareto had observed that around 80% of the land in Italy and other European countries was owned by 20% of the population. Juran (1951) noted that 80% of quality problems are caused by 20% of the potential causes, and referred to this as the '*Law of the Vital Few*'.

The Pareto Principle is a 'Rule of Thumb', not an exact relationship. If we apply the principle to estimating, whether that is in relation to cost, resource, time, performance etc., the key message is that it is important to look at the main drivers, rather than try and build a model that captures every variable (*the impossible dream*). We can think of the '*vital few*' in the Pareto sense as being Primary Drivers.

We will discuss options for dealing with the 20% '*trivial many*' under Secondary Drivers later in this chapter.

As a 'Rule of Thumb' the 80:20 Rule is an over-generalisation. We will discuss this further in Volume II Chapter 4 in the context of the Pareto Distribution.

4.1.1 What's the drive behind the Drivers?

The main reason for identifying both the Primary and Secondary Drivers is so that we can use them as an integral part of the estimating methodology and techniques we have elected to use (*or in some cases have been told to use*). Typically a Top-down Approach is likely to use high level metrics based on a Quantity of something multiplied by a Value per Unit. For example:

'101 Dalmatians' at **'75 pence per Dalmatian each day to feed'** gives us an estimate of '**£530.25 per week on dog food'** unless we can 'spot' a cheaper brand. *(Sorry, I couldn't resist it – I like to think that I am a bit of a wag!)*

If we are using a Bottom-up Approach we can apply the principles at two levels. At the lower level, we can examine the Primary Drivers for the detailed tasks or activities and use a metric-based technique as above. Alternatively, we can use the higher level drivers and associated metrics as a check function to validate or challenge the aggregation of the lower level estimates.

Typically a Top-down based estimate will have considered fewer Primary Drivers than an equivalent Bottom-up based estimate which by definition goes into much more detail, covering more discrete estimating elements by default.

So, if it were not for Pareto and Juran, we estimators may have had to worry unduly every time we didn't estimate everything in ultimate detail.

4.2 Primary Drivers

Definition 4.1 Primary Driver

A technical, physical, programmatic or transactional characteristic that either causes a major change in the value being estimated or in a major constituent element of it, or whose value itself changes correspondingly with the value being estimated, and therefore, can be used as an indicator of a change in that value.

Rather than talk generically about Primary Drivers, it is probably easier to talk about specific examples. For instance, a Primary Cost Driver would be a feature or characteristic that we might use as a good indicator of the overall cost. A Primary Schedule Driver would be one that we might use to indicate the overall duration or elapsed time for a project or product through to its completion, or to some major milestone. The important thing is that the Driver is indicative of the level or quantity we are trying to estimate; it does not have to be the cause of that effect. This simplifies matters as we do not need to get into philosophical debates about cause and effect; we need only to consider that there is a good association. This gives us more freedom to consider different indicators other than 'pure causative drivers', and allows us to give some credibility or rationale in a basis of estimate to explain why an estimate is the value we have created.

We can split these Primary Drivers into two camps:

- Those that would cause an estimated value to change (true causative drivers)
- Those that would not cause an estimated value to change but could be used to explain or mirror that change

The same entity or thing can fall into both camps simultaneously. For example:

- The weight of raw material may be a good choice of a Primary Driver for estimating the cost of raw materials
- However, weight might also be considered to be a good indicator of the labour cost (by implying more or less work content) without necessarily being the direct cause of it

4.2.1 Internal and external Drivers

To help in identifying the Primary Drivers of either sort, we might want to consider the inputs, outputs or internal transactions that might be good indicators of the activity levels in question:

- Key outputs from a process or function, e.g. design drawings issued, invoices cleared for payment, components shipped
- Key inputs to a process or function, typically from a customer, supplier or internal stakeholder, e.g. design drawings released, invoices cleared for payment, components ordered, components rejected. (*If you are getting that feeling of déjà vu, it's a similar list to the above; let's not forget that outputs from one area may be input drivers to another area, hence the similarity in the lists*)
- Key internal transactions within a process or function that reflect the activity throughput e.g. the number of software system interfaces tested, software lines of code produced (*SLOC*). Note: SLOC are not an output *per se*, but a means of expressing the level of activity in producing an end result, which would be a computer program or module

These Drivers may take many forms but are usually associated with things that we can measure or count. In general, we may consider these to be the Primary Drivers.

External drivers are those which do not directly reflect the detail level of activity within the product, process or function being estimated. However, they can be used if we can demonstrate that there is a reasonable correlation between the external driver and the activity being estimated. We might consider these external indirect drivers to be Secondary Drivers ...

4.3 Secondary Drivers

Definition 4.2 Secondary Driver

A technical, physical, programmatic or transactional characteristic that either causes a minor change in the value being estimated or in a constituent element of it, or whose value itself changes correspondingly with the value being estimated and can be used as an indicator of a subtle change in that value.

Secondary Drivers are those Drivers for the '*trivial many*' in a Pareto sense. (*Some might say they are more like passengers than drivers.*) They would cover both of these situations:

*   Minor elements of the estimate not covered by the Primary Drivers (*filling the gaps*)
*   Characteristics that influence subtle variations in major elements of the estimate already covered by the Primary Drivers (*or dotting the i's and crossing the t's*)

The latter will be dealt with more holistically in Volume III on Best Fit Lines, Curves and Mathe-Magical Transformations. In terms of the former, those elements of the estimate not covered by Primary Drivers, we have two principle techniques we can use:

1. Identify a separate metric and driver for each element of the estimate, or for a group of them, or for all of them together
2. Roll up the residual elements into the metric of one or more of the Primary Drivers

We might think that a variation on the theme would be a halfway house solution, in which we identify a factor metric that relates the minor elements of the estimate back to the Primary Driver elements, which is sort of applying option 1 to get option 2. If this aids transparency within the basis of estimate, then it is to be encouraged, but in reality it is not a separate option.

The thought of applying a single imperfect driver to a group of unrelated cost elements may seem disturbing to some, but consider it from the perspective that we are looking at the '*trivial many*' in Pareto terms. Any inaccuracy in the driver metrics will be lessened by its contribution overall. For example, consider the case in which the uncertainty around the driver metric being used is considered to be $\pm50\%$, and the driver is being applied to 20% of the overall value, then the net contribution to the overall estimate's uncertainty will $\pm10\%$.

Examples of the above are shown in Table 4.1.

Note that values may not be exactly the same when we use the alternative techniques. These are estimates, and in reality, the outturn may be close to these, but not necessarily the exact value of either of them.

Table 4.1 Alternative Techniques for Dealing with Secondary Drivers

Option	Cost Element	Driver	Quantity	Metric	Units	Total Hours	Contribution	Notes
Option 1	Manufacturing	Number of Parts	1000	1.3	*Hours/Part*	1,300.00	96%	Primary Driver
(Full breakdown)	Inspection	Parts Inspected (1 in 10 parts)	100	0.125	*Hours/Inspection*	12.50	1%	Secondary Driver
	Quality Control	QC Reports (1 in 4 Inspections)	25	1.7	*Hours/Report*	42.50	3%	Secondary Driver
						1,355.00		
Option 1	Manufacturing	Number of Parts	1000	1.3	*Hours/Part*	1,300.00	96%	Primary Driver
(Grouped)	Inspection/QC	Parts Inspected (1 in 10 parts)	100	0.55	*Hours/Inspection*	55.00	4%	Secondary Driver
						1,355.00		
Option 2	Manufacturing	Number of Parts	1000	1.352	*Hours/Part*	1,352.00	100%	Primary Driver
(Roll-up)						**1,352.00**		
Option 2	Manufacturing	Number of Parts	1000	1.3	*Hours/Part*	1,300.00	96%	Primary Driver
(Factor Roll-up)	Inspection/QC	Uplift Factor	1300	4.0%	*of Manuf Hours*	52.00	4%	Secondary Driver
						1,352.00		

4.4 Practical issues with Drivers

Whenever we identify a potential Driver of cost, schedule etc., we need to stand back and ask ourselves the pragmatic question, '*Is it useful?*' For a Driver to be useful it has to be something that is either provided to us as a statement of fact or as an assumption for the case in question; consequently, it has to be something that we can retrieve, estimate or predict in advance. If we cannot count, measure, categorise or predict the quantities or volumes of the Drivers identified, then (pragmatically speaking) we cannot use them in an Estimating Relationship. Just because we can count things after they have occurred and use them to derive Estimating Metrics does not mean we can use them as Drivers, if we cannot sensibly predict them as inputs to the estimating process.

Let's consider a number of scenarios. It is always good practice, wherever possible, to plot the data we have in order to convince ourselves just as much as anyone else. In this case, the most useful plot is a scatter diagram showing Actual Value (y) versus Driver Quantity (x):

i.   Single Primary Driver, Single Actual Observation

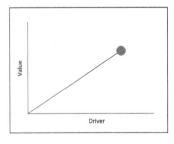

Here, we have the simplest of all Estimating Relationships from which we can derive a simple Metric to get an 'Actual Value per Driver Unit'.

Not very convincing as an estimating relationship, but in some cases it may be all we have to go on. Note, however, before you feel inclined to dismiss it that this is precisely what we are implying when we draw an analogy with something else.

By implication, we are drawing a straight line through the origin (see Chapter 2 Section 2.4.1).

ii. Single Primary Driver, Multiple Actual Observations

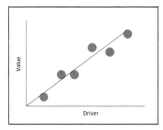

If we observe a pattern that is *'broadly speaking'* a straight line that passes through the origin, then we can surmise that there is good evidence that the driver is a Primary one. The slope of the line implies a simple estimating metric of an 'Average Value per Driver Unit'. This is the parametric equivalent of the first case, and visually is more convincing.

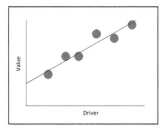

We may, however, find an equally compelling relationship implied by a straight line that does not pass through the origin. In this case we cannot derive a simple metric but we can derive a simple parametric linear relationship (see Volume III Chapter 4). It could also be a local approximation to a curved relationship (see below).

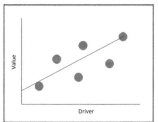

We may find that in either of the above cases there appears *'loosely speaking'* to be a straight line, i.e. more scattered than above. This may be indicative of there being other Primary or Secondary Drivers which we have not identified. We will discuss such situations under heading iii below.

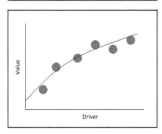

Alternatively, and perhaps more frequently, we may observe that the data follows a pattern that is more curved rather than having a straight line. We will consider these relationships more in Volume III Chapters 5 and 6.

iii. Multiple Primary Drivers, more Actual Observations than Drivers

Whilst it is possible to draw a 3-D graph with two Drivers against a set of observed values, it is often not very clear what is happening and which (if any) driver is the more dominant. To overcome this, we can plot scatter graphs of each potential Driver in turn against the observed values as above. Those graphs which show a closer relationship (straight line or curve) are potentially the Primary Drivers. Plots that are more randomly scattered are probably more indicative of Secondary Drivers. However, we should be wary of interdependence amongst drivers where two drivers are in effect linked and are measuring the same thing. Clearly, we only need one of them.

Also, in many cases, especially where there are several potential Drivers, the Drivers will each contribute towards the observed values and in effect dilute the impact of other Drivers. We will consider alternative more structured ways of identifying and calibrating drivers in Volume III on Best Fit Lines, Curves, and Mathe-Magical Transformations and the parametric relationships that they create.

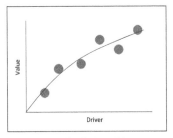

Strong Relationship

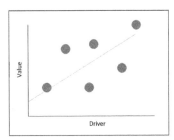

Weak Relationship

iv. Multiple Primary Drivers but with less Actual Observations than Drivers

If we have less actual values than we have potential Drivers, then we must eliminate some of the potential drivers. To some extent, we can follow the procedure above in heading iii and plot each driver in turn against the actual values, or we can jump forward to Volume III for a more statistically stringent approach.

For the Formula-phobes: Why do we need more data than Drivers?

We need two points to draw a perfect straight line. If we only have one point on 2-D surface (e.g. paper), we can draw an infinite number of lines through that point. For instance, if we lay a pencil flat on the paper and hold it fixed at one point; we can still rotate the pencil around the fixed point.

Similarly, we need three points to draw a perfect plane in 3-D space (*provided that they are not all in a straight line, of course*). Consider a piece of card held up in the air with two diagonal corners fixed in position by our finger tips. We can hold the piece of card steady at these two places in space, but we will still be able to rotate the card with our thumbs about the line formed by the two points. Fixing the third point in space with the tip of our nose, stops the rotation.

Physical analogies of 4-D surfaces and more are beyond my metaphysically challenged brain; suffice it to say that the number of dimensions defines the minimum number of points required to draw a perfect fixed multi-dimensional shape. Each dimension represents one Driver.

If we have more points than we have dimensions we have to minimise the error between the points and the 'shape' – we call this the 'best fit' line or curve, or relationship

4.4.1 Sub-classification of Primary Drivers

There will be occasions where we might look at a scatter plot and think that there may be a better way than defining a single metric for a Primary Driver, or to specifying a parametric equation based on the best fit straight line or curve through some data. We could instead choose to define a number of discrete metrics for each Driver based on a sub-categorisation or classification of the observed items.

The Primary Drivers do not have to be simple entities, such as overall parts count, we can define them in relation to other features or characteristics (i.e. parameters) that define or convey size, complexity or variety – in other words a form of classification, which may be factual (quantitative), or which may include our (*or someone else's*) qualitative assessment. For example:

- Outputs from a process or function may be sub-divided to convey complexity or effort required, e.g. the number of drawings by drawing type (*detail drawings vs. assembly drawings*)
- The physical size or properties of items may suggest clusters of data that could indicate that items may consume more of less resource depending on some broad groupings e.g. Size (small, medium, large), Material Type (metallic, mon-metallic), Physical Form (2-D Flat, 3-D Curved Surface)
- The maturity of the technology used might be a consideration and could be classified by date of introduction or perhaps Technology Readiness Levels (Mankins, 1995, and ASD(R&E), 2011). Note: Date is not a perfect measure of technology used, but could be indicative in a competitive environment. It may be considered to be a Secondary Driver in other instances
- Environmental conditions may also be an appropriate Driver. For example, the cost or time to perform maintenance and repair activities may be significantly affected by the Working Environment, and Facilities available, e.g. indoor/outdoor working, accessibility etc.

Figure 4.1 illustrates such an approach where sub-categories of Small, Medium and Large Machined parts have been identified. For each group of parts, the Manufacturing Lead-time has been based on the average of all parts in those groups.

Taking the average value for each classification is not the only choice we have; we may choose to use one of the other 5-Ms (Minimum, Mode, Median, Mean, Maximum) which are discussed further in Volume II Chapter 2 on Measures of Central Tendency.

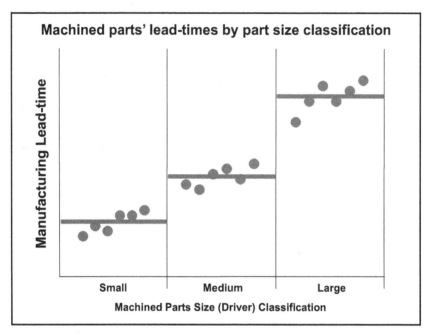

Figure 4.1 Example of Multiple Metrics for Classification Drivers

4.4.2 Avoid pseudo-drivers

Amongst my '*pet hates*' in the world of estimating are those activities that are based on a 'Level of Resource' statement; too often they ask more questions than they answer. For example, an activity or task described as a 'Two Person Job for Three Months' may imply a cost level but prompts several questions:

- Is that the same as 'one person for six months', or 'three people for two months'?
- Is that two people, all of the time?
- Does it matter which two people we choose in terms of their competency or commitment?
- How many hours do the two people each work in a month, and is that the same every month for both of them?
- What external or internal factors would need to change to have an effect on this level of effort?
- How long is that piece of string?

In short, there is no suggestion of what is driving or mirroring the activity levels. Worst of all, they tend to become a self-fulfilling prophecy!

We will all have come across situations when it is difficult to articulate the output or input to a process in tangible terms, but the problem is most worrying when it makes a significant contribution in the overall context, i.e. when it is part of the Pareto 80% by value. If this is the case, then we really ought to be seeking some measure that reflects why the level is what it is. If instead it is part of the 20% by value, then we can probably consider it in the same context as Secondary Drivers.

There will, of course, be genuine cases of fixed or schedule dependent resources. These may be due to legislative safeguards (e.g. Health and Safety) or due to some chemical or physical process such as cure time or drying time, but they can be expressed in those terms.

Example of an alternative

An example of avoiding a bland 'Level of Resource' estimate such as a '*one project controller for six months*' might be to say that there are monthly review meetings with the Operations Team requiring an average of a number of hours per month to prepare and fulfil any actions. The Driver could be the number of Review Meetings and the Metric would be the Review Preparation Time and/or the Action Clearance Time. The main benefit is that it gives an opportunity to understand and potentially challenge current practice.

OK, I'll put my soapbox away again now.

Incidentally, just because I have 'pet hates', do not misread this as 'I hate pets.' I love pets; they are very therapeutic – all estimators need some physical contact with the real world!

4.4.3 Things are rarely black or white

Primary Drivers and Secondary Drivers are not absolute. They only have that differentiation in the context within which they are being used:

With a Top-down Approach to Estimating the Secondary Drivers may well include things that are considered as Primary Drivers for a detailed element of a Bottom-up Approach to estimating because of the level of granularity implied.

4.5 Accuracy and precision of Primary and Secondary Drivers

If all this talk about Primary and Secondary Drivers and the use of 'typical' values worries you, then perhaps we should stand back and think about accuracy and precision, and get them into some sort of perspective.

Accuracy and precision are words that are frequently used interchangeably in some areas of society, but to estimators and other number jugglers, they do have distinctly different connotations. For example, we can be very precise in our estimate and quote it to six decimal places – that does not mean that it is accurate as it can still be precisely inaccurate (*in other words, precisely wrong!*).

A word (or two) from the wise?

'It is better to be vaguely right than exactly wrong.'
Carveth Read
Philosopher, logician and author

However, there is another nuance on the term 'precision' that implies a level of repeatability or reproducibility, and it is this definition that more frequently gets jumbled up with 'accuracy', which implies being broadly 'on target'.

When we are developing metrics based on the primary drivers, we should heed the advice of philosopher Carveth Read (1914, p.351), and ask ourselves, '*How accurate or precise do we need to be?*' To answer that it would help if we knew what characteristics each term was trying to convey. Before we discuss them in relation to an estimate, it may be easier to consider them in relation to measurement against a known standard.

Consider the following diagrams in Figure 4.2. If the centre of the bullseye represents the standard against which we are being measured, then pictorially we can represent a low and a high degree of accuracy against a low and a high degree of precision.

The Utopian solution would be for data to be 'Precisely Accurate', but this is all too often an unrealistic expectation. However, we can say unequivocally that we do not want our data to be 'Imprecisely Inaccurate' (*or to put it another way 'roughly random'*).

So, if we cannot have perfection, and we don't want 'random', which perspective should we settle for: 'Precisely Inaccurate' or 'Imprecisely Accurate'? In general, either might be acceptable us:

- We can certainly tolerate the variability of the 'Imprecisely Accurate' because it is telling us that on average we will be accurate. (Of course, if we look at it another way we might conclude that it is telling us '*We win some, we lose some*')
- To be 'Precisely Inaccurate' should be acceptable also if we can measure how inaccurate we are in relation to the bullseye or the standard and make an appropriate adjustment; this is akin to making an adjustment for performance against a standard

From that we can probably agree that accuracy and precision are not the same thing, although they are related, in that they both relate to a set of data measured against a defined standard or expectation. In terms of estimating, that expectation is that the estimate reflects the eventual outturn. To some degree even though we could define the degree of accuracy and of precision quantitatively, there will always be that element of qualitative or subjective opinion as to what is acceptable in different circumstances and to different people.

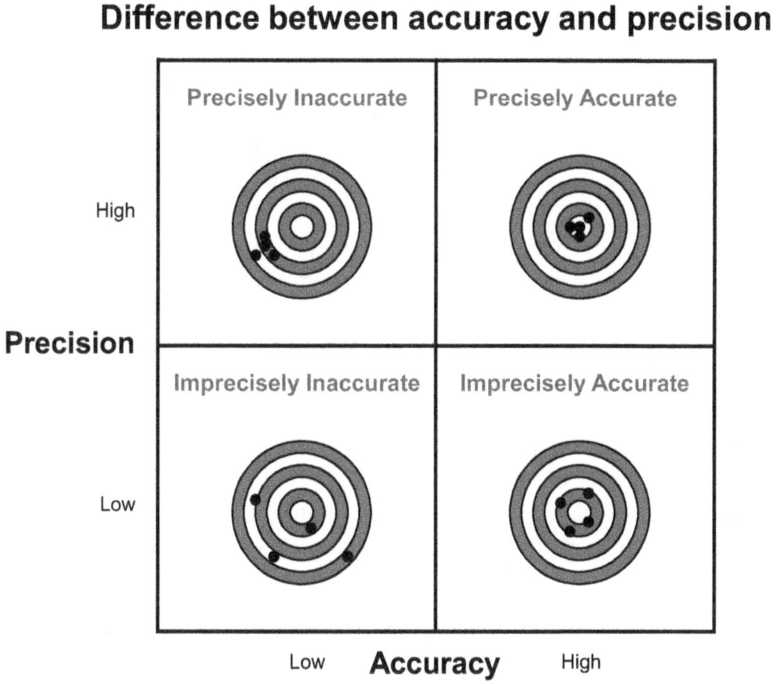

Difference between accuracy and precision

Figure 4.2 Difference Between Accuracy and Precision

Definition 4.3 Accuracy

Accuracy is an expression of how close a measurement, statistic or estimate is to the true value, or to a defined standard.

Bizarrely, the definition of precise is somewhat less precise. We have two distinct uses of the term 'precision' in estimating:

Definition 4.4 Precision

(1) Precision is an expression of how close repeated trials or measurements are to each other.
(2) Precision is an expression of the level of exactness reported in a measurement, statistic or estimate.

In the context of estimating overall, we might want to think about Estimating Accuracy as a measure of something being representative of the eventual outturn value against an agreed set of assumptions. (*When did assumptions ever 'never change'?*) Furthermore, a range estimate has a better chance of being right than a point estimate in relation to the eventual outturn, so we need to note the range or the scatter around the average. (*If that sounds like I'm saying we should hedge our bets, then 'Yes', it is more informative for the business purpose.*)

Estimating Precision (as per definition 1) might be considered as a measure of consistency in producing a similar value by different estimators' interpretation of input assumptions.

- Standards or Norms, especially in machine intensive operations may be considered to be precise, producing consistent results for the same defined scope of work, which typically would be quite specific. However, they may not be accurate as they do not necessarily consider unplanned activities.
- Another case, on a more macro scale of multiple trials leading to a consistent (precise) result can be seen through Monte Carlo Simulation. (We will revisit the accuracy vs. precision debate in this context in Volume V Chapter 3.)

4.5.1 Accuracy, precision and Drivers – A Pareto perspective

We might like to think of the difference between Primary and Secondary Drivers from a Pareto perspective:

- Primary Drivers are equivalent to the '*vital few*' in Juran's definition of Pareto and can be aligned with the determining the accuracy of the estimate.
- Secondary Drivers are akin to the '*trivial many*' and can be considered to be synonymous with refining precision from an exactness point of view.

4.5.2 Cone of Uncertainty

When it comes to accuracy and precision, there is a clear link to 'uncertainty' in an estimate, and we must set our expectations according to what might be a reasonable level based on the information available at a particular point in the lifecycle of a project, product or service. Boehm's Cone of Uncertainty (1981) famously (*or was that infamously?*) sums up the issues succinctly in relation to software estimates. However, this basic concept was an extension of earlier observations by Bauman in 1958, drawn from the chemical industry, Bauman observed differences at the conceptual stage in the region of 50% to 200% of the final outturn, whereas Boehm noted a range between 25% and 400% in software projects. Others have observed similar 'funnel curves' but to varying degrees of uncertainty.

Figure 4.3 is an adaptation of the principles expounded by Bauman, Boehm and others, reflecting the concept of accuracy improvement through the life of a project and

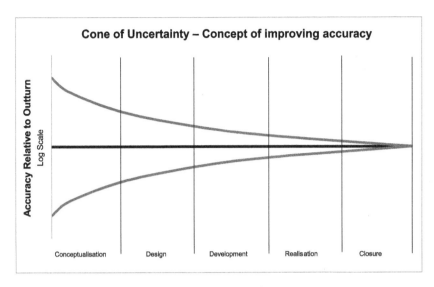

Figure 4.3 Cone of Uncertainty – Concept of Improving Accuracy

estimate. Improvement Rates across the Life Cycle Phases are illustrative only and will vary across industries, organisations and entity being estimated (i.e. cost, schedule duration etc.), and might be considered to be more extreme than may be the case in many situations, which may experience narrower Cones of Uncertainty.

Figure 4.4 illustrates the equivalent cone for an Estimating Accuracy Range advised in the AACE International Recommended Practice No 18R-97 (Dysert, 2016). Furthermore, whilst in a mathematical/scientific sense the term 'Order of Magnitude' is often used in relation to an expression of size, typically in powers of ten, but in practice the level of accuracy intended by those requesting a ROM estimate is probably much lower than a factor of ten. Compare this with Boehm's Cone of Uncertainty which is only a factor of four and is probably considered to be a Very Rough Order of Magnitude (VROM).

Whilst the edges of Boehm's Cone of Uncertainty (25%, 400%) may be considered to be extreme, they are probably not the absolute worst case scenario either! Whilst the cone might be an honest reflection of what might occur, and we are right to highlight it, it is unfortunate that the extreme width of Boehm's Cone can have a tendency to denigrate the standing of the professional estimator, fuelling the snide remarks of cynics, who might point at this as confirmatory evidence of their rhetorical question, '*Oh, it's a just guess then?*'

What can we say in our defence? The AACEI advice (Figure 4.4) recognises that on at the extreme left of the life cycle, there is virtually no definition available on which to work. It is often virtually all assumptions and guesses, not facts.

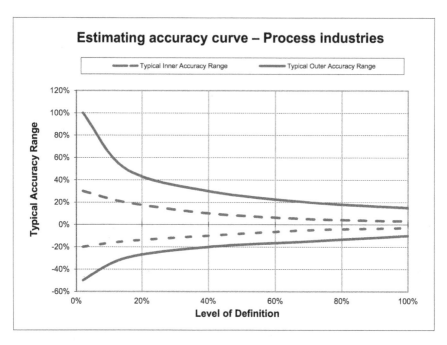

Figure 4.4 AACEI Typical Accuracy Ranges

Using an analogy from home, we wouldn't expect a plumber to repair a leaking pipe without tools and proper materials. Information is the raw material which estimators need to ply their trade; if we don't have quality materials, we cannot fix our leaks! (*Oops, out comes that soapbox again!*)

The main point of reproducing the cone of uncertainty here is to illustrate that:

- '*As a rule of thumb*' accuracy will improve over the life of an estimate
- It is better to highlight a range of values rather than express a definitive point value that may be accurate but will not be precise in terms of exactness in relation to the eventual outturn

4.6 3-Point Estimates as a measure of relative accuracy and uncertainty

Let's consider a discrete element of work.

We can use a 3-Point Estimate to express the range of uncertainty in the outcome of that element of work in terms of the time, cost or some technical parameter (such as weight) that we are estimating.

Typically, we would express this range of uncertainty using 3 values:

1. An **Optimistic Value**: one which we might achieve if everything falls neatly into place, but that we are unlikely, realistically, to better. This is sometimes referred to as the Minimum, but may not be in an absolute sense

2. The **Most Likely Value** that we will achieve. This would often be the single point deterministic estimate value that an organisation may choose to develop by a Bottom-up Approach. However, there will be occasions when the 'middle' or second point is expressed as an Average value or sometimes the Median Value depending on the circumstances. The differences between these will be discussed in Volume II Chapter 2, Measures of Central Tendency

3. A **Pessimistic Value**: one at which we might outturn if we do not perform as well as we expect or to which we aspire on the element of work. This is sometimes referred to as the Maximum, but it may not necessarily be the true absolute maximum

Now I can see that some of us may be thinking, '*Typical estimator, hedging as usual! Why can't they get off the fence?*' Well, to put it simply, it's a question of responsible professionalism. As estimating is not an exact science, there is a growing acceptance that a range estimate expressed as three points is good practice, as it facilitates a better appreciation of the issues that should be considered for certain critical business decisions. For example, making business pricing decisions on Most Likely Values alone will have a detrimental effect on the margin as we are more likely to overspend the Most Likely value than we are to underspend due to the natural positive skewness of cost distributions. (We discuss this in Volume II.)

If you were the CEO of your organisation, which would you prefer:

- A smaller chance of achieving the required margin based on a more competitive price
- Or, a better chance of achieving a margin based on a more realistic price?

3-Point Estimates help to facilitate those types of decisions, especially when the three points are linked to a level of confidence or probability. (Again we will discuss these concepts more fully in Volume II.)

4.7 Precision as an expression of appropriate or inappropriate exactness

A question that vexes some estimators is 'rounding'. At what level of exactness or precision should we cite our estimates? The

> ### A word (or two) from the wise?
>
> *Augustine's Law of Definitive Imprecision*: '*The weaker the data available upon which to base one's conclusion, the greater the precision which should be quoted in order to give the data authenticity.*'
>
> **Norman R Augustine**
> Law No XXXV
> 1997

answer is probably 'less than many of us do' (*myself included*). Norman Augustine (1997, p.231) recognised that greater precision in the reported data does not imply greater accuracy, but it can create a misplaced illusion of rigour.

There are a number of factors that should influence the decision on the level of precision required:

- The purpose of the estimate
  - If we are being asked to create a Rough Order of Magnitude (ROM) or Ballpark Estimate, then by implication we are not looking for precision
- The precision with which data is held in the reporting systems used to measure performance within the organisation
  - If the data collection systems only hold two decimal places, why produce estimates to a greater precision?
- The precision in which key inputs to the estimating methodology to be used are provided by a third party within the organisation
  - If the data has been rounded on input (say, an integer input has been rounded to the nearest ten) then this will ripple through all the calculation levels

On the other hand, there is a sound argument that if numbers are input to the chosen estimating methodology at a level of exactness that is recognised (even if the relevance of that precision is not understood by those providing it), then accept them and use them. From a practical point of view, we are probably better allowing the calculations to flow through naturally through the various steps and applying rounding to the answer as the last step; that way any rounding errors are much easier to locate, and the audit trail from

---

Rounding in Microsoft Excel

In Microsoft Excel, there is a round function **ROUND(number, num_digits)** which allows you to round a **number** to a specified number of digits, **num_digits.** Most people use this to round to 0, 1, 2 or more decimal places.

Did you know that by using negative values for **num_digits** parameter you can round a number to the nearest power of ten?

For example, **ROUND**(123456.789, 2) = 123456.79 (rounded to two decimal places)

ROUND(123456.789, -1) = 123460 (rounded to the nearest ten)

ROUND(123456.789, -2) = 123500 (rounded to the nearest hundred)

ROUND(123456.789, -3) = 123000 (rounded to the nearest thousand)

In other words, there's no need to start all that palaver of 'dividing by ten, rounding and multiplying by ten again' etc. just to round to the nearest ten.

input to output is simplified and is more transparent. We are less likely to be challenged by '*That's not the number I gave you!*'

If we are citing a range estimate (3-Point Estimate), which is recognised by many to be good practice, then we are making a statement that the eventual outturn could be any value in that range. There is a conspicuous absence of exactness there. We should resist the temptation to cite 3-Point Estimates without first rounding the Optimistic (Minimum), Most Likely and Pessimistic (Maximum) values in the range. The level of rounding will really depend on the culture of the organisation requesting or receiving the estimate.

4.8 Chapter review

The Pareto Rule is a key 'Rule of Thumb' for estimators. We should concentrate our efforts on the more significant elements of an Estimate, which in general means identifying the Primary Estimating Drivers (or the '*Vital Few*') as opposed to spreading our time and efforts unnecessarily across all constituent elements of an estimate including its Secondary Drivers (or the '*Trivial Many*').

By concentrating on the major elements and drivers we are, in essence, driving towards a reasonable level of accuracy rather than a level of precision (exactness) which is probably unnecessary, inappropriate and unattainable in reality.

It is the other definition of Precision that is important; the one that refers to repeatability. The rationale and logic used should be clear, and be capable of being reproduced with the same source information by the same estimator at a different point in time, or by another estimator at any point in time. If the source information changes, then the impact of the change on the Estimate should be evident to all.

In identifying the Primary Drivers of an Estimate, the estimator is encouraged to talk to the technical experts in the area or field concerned. To be useful, Drivers must be countable or measurable, and predictable where they are not already given or assumed within the agreement of the Estimating Scope. Useful measures to consider would be Process or v inputs or outputs, or potentially internal transactions. Other tangible measures that may be suitable as drivers include technical and programmatic characteristics.

We should also keep an open mind about using classification variables that differentiate between the likely estimate values for different sub-groups. Examples do not need to be too sophisticated, so classifications such as 'Small', 'Medium' and 'Large' would not be inappropriate so long as there is consistency used in the definition and interpretation (if this is not the case then we are running the risk of failing the repeatability principle). However, it is of paramount importance that we define the boundaries of what constitutes 'small, medium and large' quantitatively, as it may differ greatly from another person's perspective.

As a generality, Primary Drivers help us to focus on an Accurate Estimate, the addition of Secondary Drivers helps us to improve the Precision.

References

ASD(R&E) (2011) *Technology Readiness Assessment (TRA) Guidance*, US Department of Defense.

Augustine, NR (1997) *Augustine's Laws (6th Edition)*, Reston, American Institute of Aeronautics and Astronautics, Inc.

Bauman, HC (1958) 'Accuracy considerations for capital cost estimation', *Industrial & Engineering Chemistry*, April.

Boehm, B (1981) *Software Engineering Economics*, Upper Saddle River, Prentice-Hall.

Dysert, LR (2016) 'Cost estimate classification systems – As applied in engineering, procurement, and construction for the process industries', *Recommended Practice No. 18R-97*, AACE International, Revised 1 March.

Juran, JM (1951) *Quality Control Handbook*, New York, McGraw-Hill.

Mankins, JC (1995) *Technology Readiness Level: A White Paper*, NASA.

Pareto, V (1906) *Manuale di economia politica*, Milan, Società Editrice Libreria.

Read, C (1914) *Logic: Deductive and Inductive*, London, Simpkin, Marshall, Hamilton, Kent & Co. Ltd.

Figure 4.4 is a re-production of Figure 1 from **AACE International Recommended Practice 18R-97**, *Cost Estimate Classification System: As Applied in Engineering, Procurement, and Construction for the Process Industries*, as revised March 1, 2016; lead author, Larry R. Dysert. Figure 1 is reprinted with the permission of AACE International, 1265 Suncrest Towne Centre Dr., Morgantown, WV 26505 USA. Phone 304–296–8444. Internet: http://web.aacei.org

E-mail: info@aacei.org Copyright © 2016 by AACE International; all rights reserved.

5 | Factors, Rates, Ratios and estimating by analogy

In this chapter, we will define what we mean by Factors, Rates and Ratios, and how we might use these to create estimates by analogy.

5.1 Estimating Metrics

Estimating requires a clear, unambiguous statement of what is in scope, and what is not. So, perhaps we should begin by clarifying that this section's title relates to using Metrics in Estimating, and not about the means of creating estimates for such metrics, which is covered implicitly by Volume II Chapters 2 and 3 on Measures of Central Tendency, Dispersion and Shape.

In estimating we frequently talk about 'Metrics'. The term 'Metric' in a business sense is a statistic that measures an output of a process or a relationship between a variable and another variable or some reference point. From that we can interpret what we might mean by an 'Estimating Metric'.

On one level we might think of it as being a measure of how good the estimate was in relation to what eventually happened. However, we can also look at it from the perspective of it being a measure of a relationship between two things, such as cause and effect, or in a looser sense, something that can be used to measure the extent to which something else changes. It is the latter two of these which we will use here, but it is important that we also take cognisance of the former. If we are to accept Augustine's tongue-in-cheek law (Augustine, 1997, p.152), which was based on his experiences in business, then we

> ## A word (or two) from the wise?
>
> *'Augustine's Law of Unmitigated Optimism*: Any task can be completed in only one-third more time than is currently estimated.'
>
> **Norman R Augustine**
> Law No XXIII
> 1997

would be negligent not to take on board the metric it implies about the potential schedule slippage!

However, if we take this to its illogical conclusion, this factored outturn itself becomes the current estimate, which should also be factored up if we accept the law. That leads us into a never-ending spiral, increasing exponentially.

More seriously, the point here being is that we should always look at the context of any metric we use. In the case of Augustine's Law of Unmitigated Optimism, the implication may be that there will always be some risk that we have not foreseen, or that has not been adequately mitigated in our plans This is a point that is also reflected in Donald Rumsfeld's famous quotation on 'Unknown Unknowns' which we will cover in Volume V Chapters 3 and 4.

Definition 5.1 Estimating Metric

An Estimating Metric is a value or statistic that expresses a numerical relationship between a value for which an estimate is required, and a Primary or Secondary Driver (or parameter) of that value, or in relation to some fixed reference point.

In their simplest form estimating relationships can be derived simply by comparing data values for two different characteristics of the same 'entity' with one another, e.g. cost and weight of a product, or quantity produced and quantity scrapped from a manufacturing process. The resultant relationship between the two elements of data can be expressed in terms of a metric in one of three forms:

- Rate
- Factor
- Ratio

> In essence, all three are derivatives of the same basic principle, and it is the way that they are used which leads to a sometimes subtle difference in terms.

If all this sounds a little familiar, then it is because it is very strongly linked to the primary and secondary drivers that we discussed in the last chapter.

5.1.1 Where to use them

We can use Factors, Rates and Ratios as critical building blocks in creating our own estimates, or in validating or challenging estimates prepared by someone else.

Factors, Rates and Ratios are also used as a primary technique in normalising data (see Chapter 6) in which we adjust the values of one or more historical records to take account of measurable or quantifiable differences in their characteristics in relation to a defined position or standard, or in relation to the project, product or service for which we wish to create an estimate.

Metrics are used when we are estimating by 'Extrapolation from Actuals' using an Analogical Method (refer to our discussion in Chapter 2). For example, we may determine that the performance against a budget or target through to the completion of a task (Estimate To Completion or ETC) will be the same as (or better or worse than) the actual cumulative performance to date.

> ### Spot the potential oxymoron!
>
> *Estimates created by someone else would be tagged as 'Trusted Source' using our nomenclature of Estimating Methods discussed in Chapter 2.*
> *It would seem to suggest that if we are validating or challenging someone else's estimate, then perhaps we do not really trust them after all!*

For example:

Cumulative Achievement to Date	40%	
Cumulative Budget Spent to Date	44%	
Net Performance to Date	90.91%	or 10% overspent

(The reciprocal of 90.91% is 110%)

By simple Analogy the Estimate At Completion would be 10% overspent

Alternatively, if we believe that the difficulties to date have been overcome, we can take the view that the Performance going forward will be 10% better:

Cumulative Achievement to Date	40%	
Cumulative Budget Spent to Date	44%	
Net Performance to Date	90.91%	or 10% overspent
Future Performance (+10%)	100%	

By aggregation the Estimate At Completion would be just 4% overspent

5.1.2 The views of others

Before we explore what Factors, Rates and Ratios might mean to us, it would be remiss of us not to acknowledge that there are differences in interpretation and application of these terms by some learned and respected groups. For instance, the International Cost Estimating and Analysis Association (ICEAA) align these three terms as follows (ICEAA, 2009):

Rate	'Cost on Parameter'	i.e. a 'Cost per Unit of ...'
Factor	'Cost on Cost'	i.e. a 'Percentage of another Cost'
Ratio	'Parameter on Parameter'	i.e. a 'non-Cost Value per Unit of ...'

Note: ICEAA was formed by the merger in 2012 of the Society for Cost Estimating and Analysis (SCEA) and the International Society for Parametric Analysts (ISPA)

This is a very neat differentiation between the three terms, but it pre-supposes that we are all cost estimators, which may not be the case. In view of this, and the need to consider estimating and forecasting from a wider perspective, we will discuss their differences here but refrain from giving precise definitions. In terms of the following discussion, these ICEAA definitions in the main become subsets of a more general interpretation, the exception being the term Ratio.

5.1.3 Underlying Linear Relationship

We have presented this chapter in relation to an underlying presumption of a Linear Relationship. This may not in fact be correct. However, if the true relationship is a Non-linear one that can be transformed into a Linear Relationship then we can apply these same principles to the transformed data (*but we'll have to wait until Volume III Chapters 5 and 6 for that little treat*). If we are considering fairly localised data, then we can often use one or more Linear Relationships as an approximation to a curve.

5.2 Rates

Rate Metrics are expressions of how something changes in relation to some measurable Driver, attribute or parameter, and would be expressed in the form of a [Value] of one attribute per [Unit] of another attribute. Typically, these attributes or parameters will be measured in different units. They are sometimes known, or referred to, as 'Burn Rates' because they express the rate at which resources (including cost or elapsed time) are consumed, depleted or 'burnt' in relation to some Primary or Secondary Driver. However, the principle is not restricted to the consumption of resources and can be extended quite naturally to demand creation and throughput measures. If our attributes or parameters are measured in the same scalar units we may want to consider a Factor Metric instead.

For example, one of the most common Rate Metrics used in Cost Estimating is the Labour Rate which expresses a monetary value (£, € or $) per hour of labour effort expended, or some measure of output (i.e. this is the ICEAA definition discussed in Section 5.1.2). Other examples of Rate Metrics could include:

* Labour hours per Kilogram
* Software Lines of Code (SLOC) per System Interface
* Component Failures per Month
* Deliveries per Week

This illustrates the fundamental property of a Rate Metric – they convert the unit of measurement of one expression, variable or estimate into another based on the units stated or implied in the Rate Metric. For instance, we can convert a weight estimate in kilograms into a labour estimate in hours by multiplying by a Rate Metric expressed in '*hours per kilogram*', or a period of time in months into an estimate of the total number of deliveries expected by multiplying by a Rate Metric expressed as '*deliveries per month*'.

In order to use any Rate Metric of the form [Value] per [Unit] of some Parameter to generate an estimate, we simply multiply the Rate Metric by the quantity defined or estimated for the metric's Parameter. For example, Table 5.1 uses a Rate Metric to estimate the Labour Hours, to which we apply a second Rate Metric to convert the Labour Hours into an estimate of the Labour Cost.

In the majority of cases we would multiply a Driver Quantity by the Rate Metric, but depending on how the Rate Metric is expressed, we could divide the Driver Quantity by a Rate Metric.

Example of an inverted Rate Metric

Suppose we have to travel 90 miles by car, mainly on the motorway, and as a 'rule of thumb', allowing for speed limits and traffic conditions, we may know that on similar journeys we have averaged around 60 mph. To estimate the time to complete the journey we would simply divide the journey distance by the average speed to get a journey time of one and a half hours

---

In Top-down estimates we might typically use one or two high level Rate Metrics to create a single constituent element of the overall estimate. With Bottom-up estimates we may combine several such Rate Metrics in a string or series of calculations. For

Table 5.1 Example Using Two Levels of Estimating Rate Metrics

	Number of Parts		Hours per Part	Labour Hour Estimate		£ per Hour	Labour Cost Estimate
Machined Parts	511	@	3.42	1,747.62	@	50.65	£ 88,516.95
Formed Parts	267	@	1.27	339.09	@	40.44	£ 13,712.80
Operations Total	778			2,086.71			£ 102,229.75

Table 5.2 Example Using Multiple Levels of Estimating Rate Metrics

	Weight (kg)		Parts per kg	Number of Parts		Hours per Part	Labour Hours Estimate		£ per Hour	Labour Cost Estimate
Machined Parts	228.1	@	2.24	511	@	3.42	1,747.62	@	50.65	£ 88,516.95
Formed Parts	49.4	@	5.41	267	@	1.27	339.09	@	40.44	£ 13,712.80
Operations Total	277.5			778			2,086.71			£ 102,229.75

example, Table 5.2 builds on the previous example by considering an earlier step that has derived an estimate of the number of parts based on an assumed, estimated or target weight.

The Rate Metrics would typically have been derived either by analogy, parametrically or by modelling:

Analogy
Whenever we draw an analogy with something else, we are implying that the there is a value we want to estimate that varies in direct proportion to changes in the Driver.

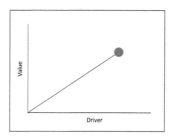

By definition, this is defining a [*Value*] per [*Driver Unit*], in other words, a Rate Metric.

Parametric
Wherever we have a number of observations which present themselves broadly speaking as a straight line passing through the origin, then we have a parametric relationship in which the slope of the line is a Rate Metric of the form [*Value*] per [*Driver Unit*].

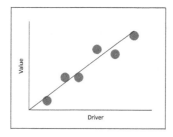

Modelled Values   Bizarre though it may seem to some, we can create estimates for Rate Metrics by simply applying a Bottom-up Estimating Methodology; an example might be Labour Rates and/or Burden Rates. Your Finance or Accounts function may create estimates of the Labour Rates going forward by analysing the forecast expenditure by cost code and the forecast of chargeable (direct) man-hours expected to be incurred for different types of resource or work, based on an assumed throughput of work in each Rate pool:

$$\text{Direct Labour Rate} = \frac{\sum \text{Departmental Operating Budgets}}{\sum \text{Departmental Direct Hours}}$$

The basis of the Labour Rates will be decided by the Finance and Accounting Policy of either the organisation or the customer in some cases.

Note: As Labour Rates are often estimates themselves, we should encourage accountants to use Monte Carlo Simulation techniques more to generate a range of values for these rates, to take account of the uncertainty around the levels of expenditure and forecast labour hours, rather than running digital scenarios which invariably will change in reality.

In Chapter 4 on Primary and Secondary Drivers; Precision and Accuracy, we introduced the idea that we could have different sub-classifications of Primary Drivers (e.g. small, medium and large parts, or, simple and complex tasks) to get around the need to determine a parametric curve. The implication, if we elect to do this, is to identify separate Rate Metrics for the different subcategories, each of which will pass through the origin, but will have a predefined limit or range of applicability as illustrated in Figure 5.1. These range limits may or may not overlap with other subcategories, depending on whether the differences in the category 'bandwidth' and their associated rates are increasing or decreasing.

The example shown in Figure 5.1 is for illustration purposes only to demonstrate the principle. If we really did have this much data at our disposal, and it were exhibiting a strong trend or pattern of behaviour such as this, then we really should be looking at a Curve Fitting technique (see Volume III Chapters 4 to 7) to identify the underlying parametric relationship. In this case, the segmented lines

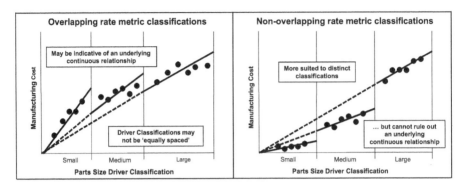

Figure 5.1 Overlapping and Non-Overlapping Classification Rate Metrics

are not always the best fit lines locally to that segment – the best fit is unlikely to pass through the origin. However, we could extend this segmentation technique to calculate a simple Rate Metric Range to quantify the uncertainty inherent in this approximation technique (see Figure 5.2), a technique we could use also for any unsegmented, simple Rate Metric.

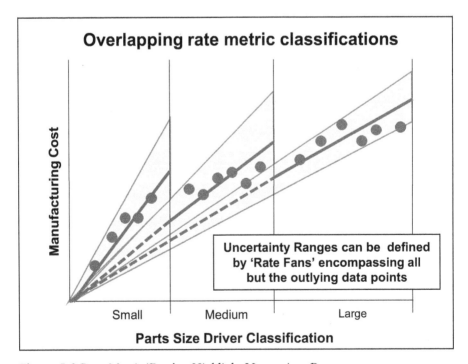

Figure 5.2 Rate Metric 'Fans' to Highlight Uncertainty Ranges

We will revisit the topic of sensitivity later in Section 5.6.

Alternatively, as we suggested in Chapter 2, we can make and document the assumption that the underlying relationship is non-linear but that it follows one of the common patterns (Exponential, Logarithmic or Power functions) that we will be discussing in Volume III Chapters 5 and 6. In this case rather than passing through the origin, the analogical assumption is that is passes through the points $(1,0)$, $(0,1)$ or $(1,1)$ respectively for the functions cited. (We will expand on this in Volume III.)

5.3 Factors

In estimating we can use Factor Metrics to express one value as a percentage of another value. By implication, as percentages, Factor Metrics are dimensionless. However, as such, when they are multiplied by another value which has an associated unit of measurement, the resulting product (*in the multiplication sense of the word*) retains the same unit of measurement e.g. £, €, $, Metres, Days, Man-hours etc. In that way they are similar to Rate Metrics with the exception that the measurement scales of the two attributes or parameters are the same.

Example of a Factor Metric

Rule 91 of the Highway Code for England, Scotland and Wales (Department for Transport, 2015) recommends '*a minimum break of at least 15 minutes after every two hours of driving.*

This can be interpreted as a Factor Metric in order to convert a Driving Time into a Journey Time (i.e. including Rest Breaks) of 1.125 or 112.5%.

$$[\textbf{\textit{Journey Time}}] = 1.125 \times [\textbf{\textit{Driving Time}}]$$

There are two interlinked ways of using Factor Metrics:

a) We can express something as a percentage of something else to create an estimate for a new element (i.e. an extrinsic factor that excludes the element to which it is applied)

b) We can express a summary estimate for something to be an uplift to another constituent element of the estimate (i.e. an intrinsic factor that includes the element to which it is applied)

Intrinsic Factor Metric = 100% + Extrinsic Factor Metric

In the journey time example, the Rest Break is 12.5% (*extrinsic*) of the Driving Time, but the Total Journey Time is 112.5% (*intrinsic*) of the Driving Time.

It is strongly recommended that Factor Metrics derived by Analogy are only used to estimate smaller elements based on larger elements rather than the other way around. Higher factors would imply that the relationship constitutes a Primary Driver, whilst smaller factors would imply a Secondary Driver relationship (see Chapter 4). Clearly, if we have derived our Factor Metric by a Parametric Method, and the relationship is sound (see Volume III Chapters 3 to 6 on Linear and Non-linear Regression), then there is no reason why we should not use it.

Example of **inappropriate** use (*just because you can, does not mean you should!*):

> Suppose Engineering Support = 25% of Manufacturing Effort
> Engineering Support Estimate = 1000 hours
> => Manufacturing Effort = 4000 hours ✘

However, there are instances where custom and practice, based on prior research, has demonstrated that Factors can be applied to smaller elements of cost to estimate larger ones. An example of this would be Lang Factors (Lang, 1947 and 1948) used to estimate the cost of installation activities in the Petrochemical Industry based on the purchase cost of equipment. It should be noted that the 'accuracy' cited is appropriately 'wide' at ±50%.

In terms of Secondary Drivers or minor constituent elements of an overall estimate then the Pareto Rule would suggest that intrinsic Factor Metrics of between 100% and 125% can be expected. However, we can only use this '*rule of thumb*' directly where we have a single Primary Driver and a corresponding single estimating element. If we have multiple estimating elements we need to check that we are comfortable that this relationship is being maintained overall.

For the Formula-phobes: Where does the 125% come from?

The Pareto Rule suggests that 80% of the effects (e.g. costs) come from 20% of the causes. We have discussed that Primary Drivers can be aligned with this 80/20 split and that Secondary Drivers can be related to the balance of 20% of the effects being related to 80% of the causes, i.e. 20/80 split.

Rule of Thumb

To get a Factor Metric that uplifts the value of the Primary Driver contribution to the Total Value, we need to divide the Total Value (100%) by the Primary Driver Contribution (80%):

$$\text{Factor} = \frac{\text{Total Estimate}}{\text{Primary Driver Contribution}} = \frac{100\%}{80\%} = 125\%$$

Table 5.3 Example Using Factor Metric

Functional Resource		%	in relation to	Labour Estimate	Units	% of Total
Engineering Hours	(Non-recurring)			5,320	Hours	
Operations Hours	(Recurring)			15,880	Hours	
Primary Driver Total				21,200	Hours	80%
Project Management	(Non-recurring)	40%	of Engineering Hours	2,128	Hours	
Project Management	(Recurring)	20%	of Operations Hours	3,176	Hours	
Project Management	Total			5,304	Hours	20%
Total				26,504	Hours	100%

Table 5.3 illustrates how two such Factor Metrics might be used to generate estimates for Project Management Hours based on a relationship with Labour Hour Estimates for both Engineering and Operations. Here, we have one Factor Metric that exceeds 25%, but is only applied to a proportion of the overall Primary Driver Hours; overall the split falls neatly within the Pareto Rule 80:20 split for Primary and Secondary Contributions. (*Don't expect all our estimates to be this perfect a fit to Pareto!*)

However, we can use one or more high level Factor Metrics as an effective 'Check Function' (rather than as a primary estimating technique in itself) to validate the relative proportions of the major elements of the overall estimate, e.g. the proportion of Non Labour Cost in relation to Labour Cost. This could be compared to a 'Rule of Thumb' metric for the organisation or industry, e.g. an organisation may typically procure 70% of the total product cost from external sources.

Such Factor Metrics will probably have been derived either by analogy (comparing it to what happened on a similar project) or parametrically (establishing a pattern of behaviour over several similar projects).

Caveat augur

It is also important to know which way round a driver works in relation to the estimate and the other drivers. Does it increase or decrease the estimate, i.e. is it directly or inversely proportional to the estimated value?

We don't want to divide inadvertently when we should be multiplying or vice versa.

In the majority of cases, as with Rate Metrics, we would multiply a reference value by a Factor Metric, but depending on how the Factor Metric is expressed we could divide the reference value by a Factor Metric. For example:

Drivers – The importance of knowing which is the right way up ...

It is probably fair to say that in respect of analogical estimating relationships most Primary and Secondary Drivers are directly proportional to the estimates they support.

Go on then, you are all going to come up with a host of drivers that are used in inverse proportion.

However (*as I was going to say*), there are other Drivers that work in inverse proportionality, i.e. as the Driver increases, the estimate reduces. There are some that work in both directions depending on what is being estimated. For example, if we were to double the number of production shifts we could double the estimated throughput (directly proportional) or we could halve the production schedule duration (inverse proportionality).

We might already have an estimate of the Standard Time Value (STV) for a particular task or activity, but we would probably want to make an adjustment to the STV estimate to take account of current or projected performance against that standard. We would do this simply by dividing the STV by the average or projected 'Performance Factor' to get a realisable time. Alternatively, we might choose to invert the performance factor and multiply it by the STV estimate. In the latter case, the reciprocal or inverse of the Performance Factor might be referred to as a 'Realisation Factor' (see Figure 5.3):

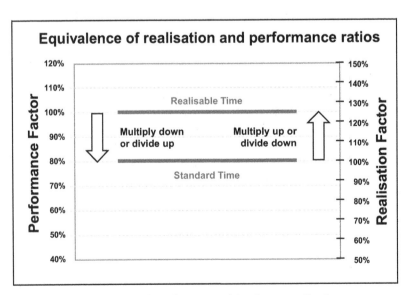

Figure 5.3 Equivalence of Realisation and Performance Ratios

e.g. **Performance Factor = 80%**

Realisation Factor = 125%

Unlike Rate Metrics, Factor Metrics maintain the unit of measurement of an expression, variable or estimate; they merely change the size or value of the entity to which it is applied.

5.4 Ratios

Ratio Metric is a quantitative relationship expressing the relative size proportions between two different instances of the same driver, attribute or characteristic. Quite frequently Ratios are expressed in terms of integer occurrences such as a 1 in 3 chance, 3 out of 4 people, and often they express a level of approximation or rough proportions. In reality, however, they can be used to express proportion between two values to any level of precision (exactness) that we care to use. Typically, they might be used to express relative differences between the corresponding attributes of two comparable products, services or other entities.

Ratios are just another way of expressing a relationship between two values that we might equally express as either a Percentage or as a Factor. For instance, if the weights of two products are 100 kg and 92 kg respectively, then we can express the ratio of their weights as:

$$\text{Ratio of Weights}: \frac{\text{Weight B}}{\text{Weight A}} = \frac{92 \text{ kg}}{100 \text{ kg}} = 0.92 = 92\%$$

(Alternatively, we can express the Ratio in the standard form often used of **n:N**. In the case above this could be expressed as **92:100** or **23:25**. However, in this context it is probably more common to express the ratio as a decimal or percentage.)

If we then use these proportions, or ratio, as an indicator of the difference in the value we want to estimate (e.g. Cost), then we are saying that the Ratio of the Costs is in the same proportion as the Ratio of the Weights:

$$\text{Ratio of Costs}: \frac{\text{Cost B}}{\text{Cost A}} = \frac{\text{Weight B}}{\text{Weight A}} = \frac{92 \text{ kg}}{100 \text{ kg}} = 92\%$$

$$\text{Cost B} = 92\% \text{ Cost A}$$

... which is directly equivalent to applying an Intrinsic Factor Metric to adjust the cost of a product by analogy with another. Consequently, as with Factor Metrics, Ratios do not change the units of measurement; they merely change the size of the value.

Within an analogical estimating technique, the value we want to estimate (e.g. cost, schedule duration etc.) is varied in proportion to the ratio of some parameter

Table 5.4 Example Using Ratios

	Product X	Product Y	Ratio (Y:X)
Weight (kg)	100	92	0.92
Complexity (relative to Product X)	100%	110%	1.10
Cost	12,500 €	12,650 €	1.012

Note: 12,650 € = 12,500 € x 0.92 x 1.10

or other. These parameters may be based on quantified differences in technical, programmatic or physical characteristics, e.g. speed, weight, build rate or dimensions, but might also be based on more qualitative or subjective parameters such as complexity. (Clearly we must express the latter against a numeric scale in order to create a numerical ratio.)

For instance, we may have concluded that the cost of a Product Y can be based on that of a similar Product X, so long as we take account of differences in the weight of the two products and the view of the technical expert that Product Y is more complex to make than Product X. We might derive a cost estimate for Product Y by analogy with Product X by applying adjustments to the known cost of Product X based on the ratio of the weights and the ratio of the difference in complexity, as illustrated in Table 5.4.

If only life was always that simple! Things do generally get more complicated when we start considering multiple Factors, Rates and Ratios.

5.5 Dealing with multiple Rates, Factors (and Ratios)

We can use multiple Rates, Factors and Ratios to create an Estimate by Analogy where we have a number of potential Primary Drivers. If we have the luxury of a number of data points which exceed the number of potential variables or drivers, then we might be better to derive an estimate by Parametric Methods that will compensate for any unintended or unforeseen interaction between the Drivers themselves.

Where we have more drivers than data points then we cannot use a Parametric Method without first differentiating the Primary Drivers from the Secondary ones – and potentially rejecting the latter.

If we have a few potential Drivers only, and even less estimate reference items, then we might want to consider generating multiple estimates by analogy based on each of the reference items in turn.

Before we get into the discussion of things we need to consider, a word of warning to avoid inadvertent but avoidable under or over estimating …

Caveat augur

In using multiple Factors, Rates or Ratios in an Analogy, it is important that the associated Drivers are wholly independent of each other otherwise they must be weighted or adjusted in some way to avoid accidental duplication or counter-action.

Ignore non-moving drivers at your peril.

5.5.1 Anomalous analogies

We might consider that the weight of a finished product is a possible indicator of the time to assembly it. We might also consider that the number of items to be assembled (including each sub-assembly as a single item) might also be a good indicator of the assembly time. However, we would probably also argue that if the weight of a product were a good indicator of its assembly time, then by the same reasoning it is also likely to be a good indicator of the number of parts. Consequently, if we were to apply Ratio or Factor Metrics for the difference in both Weight and Item Count to our Estimate Reference Point, then we run the risk of double counting, or partially double counting, as illustrated in Table 5.5.

Suppose instead that the assembly to be estimated is heavier than the Reference Assembly, but that it contains fewer, but larger and heavier items, including potentially more sub-assemblies in place of discrete detail parts. It is possible in using both Drivers that they cancel each other out as illustrated in Table 5.6. This is entirely possible, of course, but it might also be wholly inappropriate. (*There I go again, sitting on the fence – I will just have to hope I don't get too many splinters!*)

So, what do we assume, and how do we deal with them? The choices seem to be:

- The Drivers are completely independent of each other (*or near enough*)
- The Drivers are wholly dependent (correlated) on each other (*or near enough*), or their values coincidentally move in the same direction. Note: Correlation is discussed in more detail in Volume II Chapter 5 on Measures of Linear Dependence and Correlation

Table 5.5 Example of Partially Duplicated Metrics

	Analogy by Weight		Analogy by Part Count		Analogy by Weight and by Part Count		
	Weight (kg)	Assembly Time	Number of Parts	Assembly Time	Weight (kg)	Number of Parts	Assembly Time
Reference Item Data	25	75	48	75	25	48	75
New Item Data	30		64		30	64	
Ratio (New Item/Ref Item)	120%	120%	133%	133%	120%	133%	160%
New Item Time Estimate		90		100			120

Table 5.6 Example of Metrics that Cancel Each Other

	Analogy by Weight		Analogy by Part Count		Analogy by Weight and by Part Count		
	Weight (kg)	Assembly Time	Number of Parts	Assembly Time	Weight (kg)	Number of Parts	Assembly Time
Reference Item Data	25	75	48	75	25	48	75
New Item Data	30		40		30	40	
Ratio (New Item/Ref Item)	120%	120%	83%	83%	120%	83%	100%
New Item Time Estimate		90		62.5			75

- The Drivers are neither wholly dependent nor independent of each other (*i.e. they are somewhere in the middle*). We call that partially correlated and deal with what this means in Volume II

On the balance of probabilities, for the example in question, the first two are extremes, therefore, it is the last one (*and typically the most difficult to deal with*) that is most likely to be representative of our reality (Figure 5.4).

We could make some sweeping assumptions that we might *'get away'* with grouping the following together:

- Those values which are very tenuously linked (weakly correlated) can be treated as being in the same camp as those that are truly independent or uncorrelated
- Those values which are closely linked (strongly correlated) can to all intents and purposes be rolled up together and classified as duplicating each other

… we will still be left with those in the middle ground where values are partly linked and partly independent of each other (partially correlated). However, the chances are such that we will not know to what extent they overlap without more evidence (*data*). If we had more data, we could consider analysing it parametrically, looking for the pattern of behaviour (see Volume III Chapters 3 to 6).

However, at the moment we are considering the problem from the point of view of an analogy, where we only have a single reference point. So what are our options?

1. **The 'Ostrich' Option:** Pick one driver and ignore the other, wishing we had never thought of it in the first place?
2. **The 'Indefensible' Option:** Do both together on the assumption that they are independent and forget that they may duplicate or eradicate each other?
3. **The 'Fence' Option:** Do both as alternatives to get a range estimate?
4. **The '50:50' Option:** Assume that they are 50% independent and 50% co-dependent? (*… or any other proportion we care to guess at – educated or otherwise.*)
5. **The 'Health & Safety' Option:** Do both as alternatives, and also together ('50:50' Option) to get a three-point estimate? (*I.e. sit on the fence with a crash mat on either side.*)

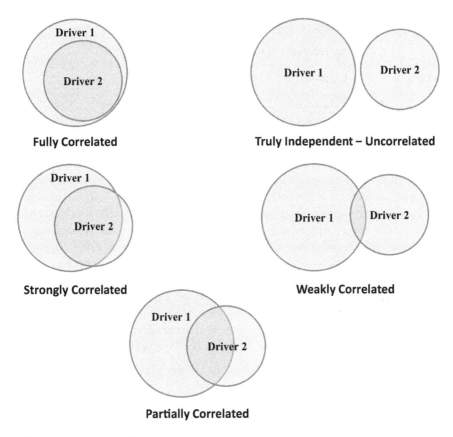

Figure 5.4 How do Your Drivers Stack Up?

Before we dive in and make a rash uninformed judgement on which to base our analogy, we should look at what that it would mean in different circumstances so that our decision is an informed one rather than an instinctive one, or (*worst still*) a random guess. Let us look at two basic estimating relationships in which the estimate is a function of two variables, Driver-A and Driver-B.

In its simplest form, the model could be either additive or multiplicative, both of which present opportunities and challenges for analogical estimates.

5.5.2 Analogies with an additive model

These are estimating models where Drivers are thought to influence the value of separate constituent elements, which we will refer to as "Estimating Elements"

Total Estimate = Sum of all the Estimating Elements

… where for each Estimating Element:

$$\text{Estimating Element} = \text{Driver Metric} \times \text{Driver Quantity}$$

A good example of such a model would be:

$$\text{Total Product Cost} = \text{Labour Cost} + \text{Non Labour Cost}$$

Here, there are two Estimating Elements, Labour Cost and Non Labour Cost. *(In many situations there are likely to be multiple Estimating Elements.)* Each Estimating Element could be the sum of other lower level Estimating Elements, but at some stage we will reach the point where the Estimating Element can be expressed as the product of a Driver Metric and a Driver Quantity.

If we apply an analogy at the constituent Estimate Element level (e.g. Labour Cost or Non Labour Cost) using a single driver for each, then the procedure is exactly as defined previously for a simple analogy. However, if we want to do this at the topmost level, we have a problem unless they have a common driver. (Rather than drop back into full mathematical notation, we will stick with the simple Labour, Non Labour example to explore the issues involved.)

Suppose that the number of items to be manufactured and assembled is considered to be a reasonable Driver of the Labour Cost, but that we do not know the Cost per Item metric. Similarly, we might also consider that the finished product weight is a good indicator of the Non Labour Cost, but again, we might not know Cost per lb or kg metric:

$$\text{Labour Cost} = \text{Cost per Item} \times \text{No Items}$$

$$\text{Non Labour Cost} = \text{Cost per kg} \times \text{Weight}$$

$$\text{Product Cost} = \text{Labour Cost} + \text{Non Labour Cost}$$

Giving:

$$\text{Product Cost} = \left(\text{Cost per Item} \times \text{No Items}\right) + \left(\text{Cost per kg} \times \text{Weight}\right)$$

Therefore, in this situation, the Analogy Ratio between Product A and Product B would be:

$$\frac{\text{Product Cost B}}{\text{Product Cost A}} = \frac{\left(\text{Cost per Item} \times \text{No Items B}\right) + \left(\text{Cost per kg} \times \text{Weight B}\right)}{\left(\text{Cost per Item} \times \text{No Items A}\right) + \left(\text{Cost per kg} \times \text{Weight A}\right)}$$

… where both 'Cost per Item' and 'Cost per kg' Rate Metrics are unknown.

With our current understanding of the variables and constants in this model, this analogy ratio cannot be simplified any further and, more importantly, it cannot be solved in general terms. However, in certain circumstances a solution is possible …

- Where Driver Ratios are equal
- Where Driver Ratios are similar

Case 1 – Where Driver Ratios are equal

Consider the particular instance when the ratio of the Number of Items between the two products is the same as the ratio of the finished product Weights. We may feel justified in concluding that the two drivers are fully or highly correlated, i.e. the Number of Items ($N\gamma$ Items) in a Product is proportional to the Weight of the Product, implying a Rate Metric of Number of Items per Unit of Weight:

$$N^0 \text{ Items} \propto \text{Weight}$$

or $$N^0 \text{ Items} = \text{Rate Metric} \times \text{Weight}$$

i.e. $$N^0 \text{ Items} = \text{Items per kg} \times \text{Weight}$$

However, the fact that the driver ratios are equal may be a just chance or fluke occurrence, and that the drivers are not correlated at all in the general sense. Nevertheless, it does not stop us exploiting the result in this particular instance, as we are using it as a one-off analogous relationship, not a parametric one. Coincidence or not, we can now simplify our analogy above to:

$$\frac{\text{Product Cost B}}{\text{Product Cost A}} = \frac{\text{Weight B}}{\text{Weight A}} = \frac{N^0 \text{ Items B}}{N^0 \text{ Items A}}$$

Table 5.7 illustrates that we do not need to know the breakdown into constituent estimating elements in order to apply an analogy where the driver ratios are equal. In the first instance (upper section), we apply the appropriate Rate Metrics to determine the Total Product Cost. In the second (lower section), we simply apply a Ratio Metric to get the same answer. (Note this only works in this instance because the Ratios of the two Constituent Driver Elements are the same.)

Table 5.7 Example of Two Constituent Estimating Element Drivers with the Same Change Ratio

Additive Model in which Constituent Element Rate Metrics are known	Labour Cost Driver	Rate Metric 1		Non Labour Cost Driver	Rate Metric 2				Implied Rate Metric 3	Implied Rate Metric 4
	Number of Items	Cost per Item	Labour Cost	Weight (kg)	Cost per kg	Labour Cost		Total Cost	Cost per Item	Cost per kg
Product A (Reference Item)	50	£25	£1,250	25	£70	£1,750		£3,000	£60	£120
Product B (New Item)	60			30						
Ratio (New Item / Ref Item)	120%	→	120%	120%	→	120%	→	120%		
New Item Cost Estimate			£1,500			£2,100	→	£3,600		

Additive Model in which Constituent Element Rate Metrics are unknown	Labour Cost Driver			Non Labour Cost Driver			Total Cost Driver		Implied Rate Metric 3	Implied Rate Metric 4
	Number of Items			Weight (kg)			Weight (kg)	Total Cost	Cost per Item	Cost per kg
Product A (Reference Item)	50			25			25	£3,000	£60	£120
Product B (New Item)	60			30			30			
Ratio (New Item / Ref Item)	120%			120%			120%	120%		
New Item Cost Estimate								£3,600		

For the Formula-philes: Rolling-up 'same value' Ratios

Consider a family of products where there are two high level Cost Estimating Elements, $C_{x,1}$ and $C_{x,2}$ where x is the product identifier

For each Product the Cost, C_x, can be expressed as:

$$C_x = C_{x,1} + C_{x,2} \quad (1)$$

Suppose each Cost Element can be estimated by the product of a Primary Cost Driver Quantity, $D_{x,i}$ and an appropriate Rate Metric, R_1

$$C_{x,1} = R_1 D_{x,1}$$

$$C_{x,2} = R_2 D_{x,2} \quad (2)$$

Substituting (2) in (1), the Cost of Product x can be expressed as:

$$C_x = R_1 D_{x,1} + R_2 D_{x,2} \quad (3)$$

Suppose that for two products A and B, the ratio of the Primary Drivers of the two Cost Elements are equal to some value k:

$$\frac{D_{A,2}}{D_{A,1}} = \frac{D_{B,2}}{D_{B,1}} = k \quad (4)$$

By re-arranging (4), we can deduce that the cross-product ratios of both Drivers must also be equal for the two products:

$$\frac{D_{B,1}}{D_{A,1}} = \frac{D_{B,2}}{D_{A,2}} \quad (5)$$

Also, by re-arranging (4) we get:

$$D_{A,2} = k D_{A,1} \quad (6)$$

$$D_{B,2} = k D_{B,1} \quad (7)$$

Substituting (6) and (7) in (3), the cost of each product can be expressed as:

$$C_A = R_1 D_{A,1} + R_2 k D_{A,1} \quad (8)$$

$$C_B = R_1 D_{B,1} + R_2 k D_{B,1} \quad (9)$$

Simplifying (8) and (9):

$$C_A = (R_1 + R_2 k) D_{A,1} \quad (10)$$

$$C_B = (R_1 + R_2 k) D_{B,1} \quad (11)$$

From (10) and (11) the ratio of the Product Costs can be expressed as:

$$\frac{C_B}{C_A} = \frac{(R_1 + R_2 k) D_{B,1}}{(R_1 + R_2 k) D_{A,1}} \quad (12)$$

Simplifying (12), for the case in question, the Product Costs can be shown to be the ratio of one of the Primary Drivers.

$$\frac{C_B}{C_A} = \frac{D_{B,1}}{D_{A,1}} \quad (13)$$

Substituting (5) in (13), we can demonstrate that the ratio of Product Costs can be determined by the ratio of either Primary Driver:

$$\frac{C_B}{C_A} = \frac{D_{B,2}}{D_{A,2}} \quad (14)$$

Where the ratio of Primary Drivers for two products are the same, the ratio of the product costs can be expressed as the cross–product ratio of either Primary Driver, allowing a simple estimate by analogy to be performed based on a Ratio Metric.

For the Formula-phobes: Rolling-up 'same value' Ratios

Suppose that we want to create an estimate by analogy for Product-B based on Product-A. Suppose also that we have previously identified that the N° of Items is a Primary Cost Driver of Labour Cost, and that Weight is a Primary Cost Driver of Non Labour Cost.

Finally, suppose also that the ratio of the N° Items for the two Products equals the ratio of the Weights for the two Products. This implies that the N° Items for Product-B can be replaced by the Weight of the Product multiplied by a Constant equal to the N° of Items for Product-A divided by the Weight of Product-A.

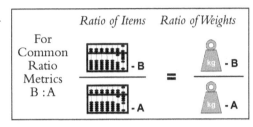

If we know the breakdown of the Constituent Cost Estimates, we can create an estimate for Product-B based on the sum of the products of the Cost Driver Quantities and their Cost Driver Metrics.

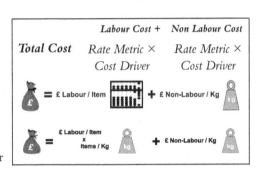

However, in this particular case we can substitute the Labour Cost Driver and Metric based on N° Items with one based on the Weight and Cost per kilogram.

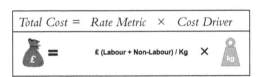

This enables us determine the Total Cost based on the ratio of the Product Weights alone without the need to know the constituent Cost Estimating Elements.

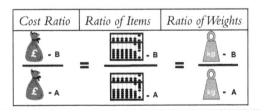

Case 2 – Where Driver Ratios are similar but not identical

If the ratios of the two sets of Drivers are not the same, but are very similar, then we might conclude that they are closely correlated (see Volume II Chapter 5.) However, as in the previous case, it could be just a coincidence which is not true in general.

$$\text{Total Product Cost} \approx \text{Weight} \times \text{Total Cost per Kg}$$

$$\text{Total Product Cost} \approx \text{No Items} \times \text{Total Cost per Item}$$

Nevertheless, in such a case we might consider using the average (Arithmetic Mean) of the two ratios. For example, if the ratio of the Product Weights is 1.14, and the ratio of the Number of Items in the Products is 1.20, we might feel comfortable in using an overall ratio of around 1.17 for the basis of the analogy, implying that the Drivers are just as likely as each other as an indicator of the overall cost. *Unfortunately, an estimator's life is not so simple.*

The Arithmetic Mean gives equal weighting to both constituent Estimating Elements, which in itself may or may not be an appropriate assumption. Suppose instead of assuming equality of driver weightings, we have a rule of thumb relationship that says that Labour and Non Labour Costs are usually in the ratio of around 1:2 or (33% and 67%) but could be anywhere in the range 1:3 (25%, 75%) and 2:3 (40%, 60%). We may want to refine our choice of overall ratio to use to reflect this by taking the weighted average of the two sets of Driver ratios. Table 5.8 illustrates the validity of this approach.

Note that any extreme difference in Ratios between the two drivers will fundamentally change the 'Rule of Thumb' relationship for the two constituent estimating elements. As a consequence, we should really consider a sensitivity analysis around the 'Rule of Thumb' weightings to assess the likely impact on the analogy in question, as illustrated in Table 5.9.

In the example, we have assumed that here are two drivers (N° of Items and Weight) whose values for Products A and B are both known. We also know the cost of Product A, and have a 'Rule of Thumb' relationship that the Labour and Non Labour Costs are in the approximate ratio range of 1:4 and 2:3 respectively. With the values cited, we can

Table 5.8 Example of Using Weighted Average Change Ratios

Additive Model in which Constituent Element Rate Metrics are known	Labour Cost Driver	Rate Metric 1		Non Labour Cost Driver	Rate Metric 2				Implied Rate Metric 3	Implied Rate Metric 4
	Number of Items	Cost per Item	Labour Cost	Weight (kg)	Cost per kg	Labour Cost	Total Cost		Cost per Item	Cost per kg
Product A (Reference Item)	50	£25	£1,250	25	£70	£1,750	£3,000		£60	£120
Product B (New Item)	60			28.5						
Ratio (New Item / Ref Item)	120%	→	120%	114%	→	114%	→	116.5%		
New Item Cost Estimate			£1,500			£1,995		£3,495		

Additive Model in which Constituent Element Rate Metrics are unknown	Labour Cost Driver			Non Labour Cost Driver			Combined Weighted Cost Driver	Total Cost	Implied Rate Metric 3	Implied Rate Metric 4
	Number of Items			Weight (kg)					Cost per Item	Cost per kg
Product A (Reference Item)	50			25				£3,000	£60	£120
Relative Driver Weighting									1	2
Product B (New Item)	60			28.5						
Relative Driver Weighting	33%			67%					Max	Min
Ratio (New Item / Ref Item)	120%			114%	↘		116%	116%	120%	114%
New Item Cost Estimate					→			£3,480	£3,600	£3,420
									Range Estimate	

Table 5.9 Example of Sensitivity Around Rule of Thumb Change Ratio Weightings

	Number of Items	Weight (kg)		Product A Cost
Product A	50	25		£3,000
Product B	60	28.5		
Ratio	120%	114%		

	Product A Input Weighting		Weighted Average Change Factor	Product B Estimated Cost	Product A Output Weighting		
	Labour Cost %	Non Labour Cost %			Labour Cost %	Non Labour Cost %	
	10%	90%	1.146		10.5%	89.5%	
	20%	80%	1.152		20.8%	79.2%	
	25%	75%	1.155	£3,465	26.0%	74.0%	
Expected Range	30%	70%	1.158	£3,474	31.1%	68.9%	Expected Range
	33.33%	66.67%	1.160	£3,480	34.5%	65.5%	
	40%	60%	1.164	£3,492	41.2%	58.8%	
	50%	50%	1.170		51.3%	48.7%	
	60%	40%	1.176		61.2%	38.8%	
	66.67%	33%	1.180		67.8%	32.2%	
	70%	30%	1.182		71.1%	28.9%	
	75%	25%	1.185		75.9%	24.1%	
	80%	20%	1.188		80.8%	19.2%	
	90%	10%	1.194		90.5%	9.5%	

determine a range estimate for the cost of Product B as highlighted. Furthermore, for the change ratios assumed, the implied weightings for Labour and Non Labour Costs for Product B are still in the range of our 'Rule of Thumb'. This will not always be the case.

Where the differences between Driver Change Ratios are greater, especially where one increases and the other decreases, as illustrated in Table 5.10, then the 'Rule of Thumb' may be challenged. In this latter case the range of outputs where the implied weighting of Labour and Non Labour Costs for Product B respect the 'Rule of Thumb'

Table 5.10 Example of Sensitivity Around Rule of Thumb Change Ratio Weightings

	Number of Items	Weight (kg)		Product A Cost
Product A	50	25		£3,000
Product B	60	20		
Ratio	120%	80%		

	Product A Input Weighting		Weighted Average Change Factor	Product B Estimated Cost	Product A Output Weighting		
	Labour Cost %	Non Labour Cost %			Labour Cost %	Non Labour Cost %	
	10%	90%	0.840		14.3%	85.7%	
	20%	80%	0.880		27.3%	72.7%	Expected Range
	25%	75%	0.900	£2,700	33.3%	66.7%	
Expected Range	30%	70%	0.920	£2,760	39.1%	60.9%	
	33.33%	66.67%	0.933	£2,800	42.9%	57.1%	
	40%	60%	0.960	£2,880	50.0%	50.0%	
	50%	50%	1.000		60.0%	40.0%	
	60%	40%	1.040		69.2%	30.8%	
	66.67%	33.33%	1.067		75.0%	25.0%	
	70%	30%	1.080		77.8%	22.2%	
	75%	25%	1.100		81.8%	18.2%	
	80%	20%	1.120		85.7%	14.3%	
	90%	10%	1.160		93.1%	6.9%	

is greatly reduced. In some more extreme scenarios, there may not be an output that follows the 'Rule of Thumb' for the inputs. In such cases, we need to challenge our basic assumptions (*if we don't, someone else will*):

- Are the Drivers the most appropriate ones for the Constituent Estimating Elements (e.g. the Labour and Non Labour Costs for the case in question)
- How valid is the expected range of the 'Rule of Thumb', after all it is only an approximate, empirical relationship by definition? How prescriptive should we be on the bounds?
- Is it appropriate to create a simple estimate by linear analogy? Suppose the cost behaviour is not linear? The 'Rule of Thumb' may work better if there is a non-linear relationship, and a more sophisticated parametric estimate may be required as illustrated in Table 5.11 using Chilton's Law (Turré, 2006) that we discussed briefly in Chapter 2

We can utilise more sophisticated models that extend the empirical relationships observed by Chilton's Law using multiplicative models, or a combination of both additive and multiplicative models.

5.5.3 Analogies with a multiplicative model

Multiplicative estimating models are those where Drivers are thought to influence the value of the Estimate overall, independently from each other. We can express this in its simplest form as the product of its drivers:

$$\text{Estimating Element} = \text{Constant} \times \text{Driver1} \times \text{Driver2}$$

Table 5.11 Example of Sensitivity Around Rule of Thumb Change Ratio Weightings with Chilton's Law

	Number of Items	Weight (kg)		Product A Cost
Product A	50	25		£3,000
Product B	60	20		
Ratio	120%	80%		
Power	0.60	0.60	=> Power Value Used in Chilton's Law	
Power Ratio	112%	87%		

	Product A Input Weighting		Weighted Average Change Factor	Product B Estimated Cost	Product A Output Weighting	
	Labour Cost %	Non Labour Cost %			Labour Cost %	Non Labour Cost %
	10%	90%	0.899		12.4%	87.6%
	20%	80%	0.923		24.2%	75.8%
	25%	75%	0.935	£2,805	29.8%	70.2%
Expected Range	30%	70%	0.947	£2,841	35.3%	64.7%	Expected Range
	33.33%	66.67%	0.955	£2,865	38.9%	61.1%
	40%	60%	0.971	£2,913	48.0%	54.0%
	50%	50%	0.995		56.1%	43.9%
	60%	40%	1.019		65.7%	34.3%
	66.67%	33.33%	1.035		71.8%	28.2%
	70%	30%	1.043		74.8%	25.2%
	75%	25%	1.055		79.3%	20.7%
	80%	20%	1.067		83.6%	16.4%
	90%	10%	1.092		92.0%	8.0%

A more holistic model based on Chilton's Law would be:

$$\text{Estimate} = \text{Constant} \times \left(\text{Driver1}\right)^{\text{Driver Metric 1}} \times \left(\text{Driver2}\right)^{\text{Driver Metric 2}}$$

The choice of model will also depend on whether the Drivers are independent of each other or whether they are partially or wholly correlated with one another. We will examine the two conditions separately.

Case 1 – Where Driver Ratios are independent of each other

An example of the former simple model might be:

$$\text{No Workstations} = \text{Cycle Time} \times \text{Output Rate}$$

This model is based on the axiomatic relationship that the number of units in operation is a function of the elapsed time (cycle time) that the units are in work, multiplied by the rate of output. The reason it is included here as an estimate relationship, even though it will always be an inviolable relationship, is that the actual cycle time and other variables, will change (through learning, variable performance and ad hoc operational problems etc.). Also, the output completion rate (or delivery rate) will vary depending on resource levels and customer or market demand. Looking forward as a planner, an estimate will be required for the initial facility set-up and the timing of additional step-up facilities. (*We could easily argue that this relationship also falls under the heading of a Parametric relationship rather than multiple analogies.*)

Now I can see what some of you are thinking – if the relationship in this example is axiomatic, as I have declared, then surely the more straightforward and logical thing to do would be to use the data assumptions directly rather than to compare it to another programme? (*By the way, you can read that as either 'Programme' as in 'Project' or as in 'Schedule' in this case.*) Naturally, you have a very valid point to argue, but consider this – what if the maximum rate does not first occur when the cycle time is at its minimum? We might only know that it is similar to Programme X, which used 12 workstations, based on a single shift pattern, but the delivery rate on the new programme is expected to be around double, and that there will be something like 25% more work content (which could translate into proportionately longer cycle time). We might also have the option of introducing a 50% Night Shift in order to reduce the cycle time. So, by analogy with Programme X, the estimated number of workstations on Programme Y might be calculated in Table 5.12.

Note that the Output Rate and Work Content Factors are Drivers that are used in direct proportion with the estimated number of workstations (i.e. as multipliers), whereas the N° of Shifts is used in indirect proportion to the answer (i.e. as a divider).

Case 2 – Where Estimating drivers are partially correlated with one another

However, we will not always be faced with the situation in which we have more than one driver that are independent of each other. Consider the case of an estimate for

Table 5.12 Example of Workstation Estimate Using Multiplicative Drivers

	Output Rate	Work Content	Shifts		Work-stations
Product X	100%	100%	1		12
Product Y	200%	125%	1.5		
Multiplying Ratio	2	1.25			
Dividing Ratio			1.5		
Net Change Ratio	Multiply	Multiply	Divide	1.67	
Product Y					20

the cost of a new ship to be created by analogy with another ship. Suppose we determine from the information available to us that the primary cost drivers are the product's gross tonnage, maximum speed and number of personnel it can accommodate (crew and non-crew). These three cost drivers are neither wholly dependent on, nor independent of, each other; reality is probably somewhere in between, but we can use the product and the Geometric Mean of the factors to calculate these two extremes:

> Gross tonnage is a measurement of a ship's volume or capacity. Also, the more personnel that a ship is designed to accommodate, then the greater is the need for private and communal space areas and support services and infrastructure. The maximum speed also affects the contour of the ship which in turn can be linked to its capacity

* We can use the product of the factors to establish a pessimistic estimate by analogy representing the case where all three drivers influence the cost independently of each other. In so doing we are inferring that:

$$Cost = Constant \times Gross\ Tonnage \times Personnel \times Maximum\ Speed$$

… which is probably an over-simplification of the true relationship

* We can use the Geometric Mean of the factors to derive an optimistic estimate by analogy that represents the case where all three drivers are largely interchangeable and the impact of any one overlaps the impact of another. As we will discuss in Volume II Chap-

> The Geometric Mean of **n** numbers takes n^{th} root of the **product** of those numbers (see Volume II Chapter 2.)

ter 2, by taking the Geometric Mean of the factors we are attempting to take a balanced view of the different cost driver factors. By implication, the cost can be expressed as being directly proportional to the Geometric Mean of its Cost Drivers in a multiplicative model:

$$\text{Cost} = \text{Constant} \times \left(\text{Gross Tonnage} \times \text{Personnel} \times \text{Maximum Speed} \right)^{1/3}$$

(The cube root here comes from the fact that we have three variable values.)

Both extremes are illustrated in Table 5.13. The difference being too extreme to ignore.

We should note that whilst the Geometric Mean offers what is probably a better technical solution to this problem, the Arithmetic Mean of the Factors will usually yield a very similar result to the Geometric Mean for values where the ratio between highest and lowest factors is less than 1.33; this is not the case for higher values, where the Geometric Mean will always be less and the difference would be more significant to the estimator (see Figure 5.5). Furthermore, as highlighted in Chapter 2, the analogical approach to estimating may not be considered appropriate for higher value change

Table 5.13 Geometric Mean in Normalising Interchangeable Factors

Cost Driver	Reference Ship	New Ship	Cost Driver Factor
Gross Tonnage (gt)	3200 gt	3600 gt	1.125
Personnel on Board	21	24	1.143
Maximum Speed	20 kt	23 kt	1.150

Product of Factors	1.479	=> Independent Linear Drivers
Geometric Mean of Factors	1.139	=> Interchangeable Drivers

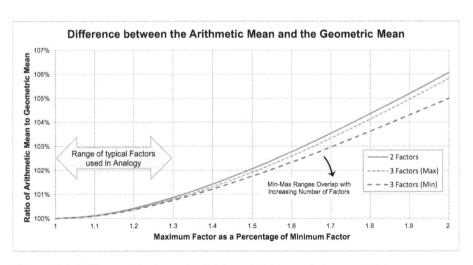

Figure 5.5 Difference Between the Arithmetic Mean and Geometric Mean

factors due to the simplicity of the underlying common assumption of a linear relationship that passes through the origin.

In terms of whether to use the Arithmetic Mean or Geometric Mean, the pragmatic question for all estimators really is *'just how precisely inaccurate do we need to be?'*

This difference is likely to be insignificant in relation to the expected accuracy range *(or inaccuracy to be more correct)* on an estimate derived in this way, especially for values at the lower end that is most likely to be pertinent to Analogical Estimating Factors. Our conclusion might be *'why bother with such an inappropriate and unnecessary level of precision provided by a Geometric Mean?'* So, do we:

- Use the Arithmetic Mean where Ratios are very similar on the basis that it is more familiar and easier, but adopt the Geometric Mean only where there is a bigger difference in our Ratios? (We would need to have some sensitivity reference table similar to Figure 5.6 and Table 5.14.)
- Use the Geometric Mean regardless as we then don't have to remember when to swap over?

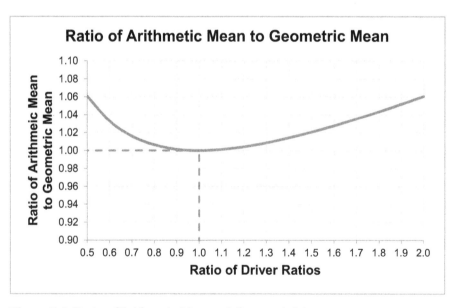

Figure 5.6 Ratio of Arithmetic Mean and Geometric Mean

Table 5.14 Differential Between Arithmetic and Geometric Means in Relation to Driver Ratios

Ratio of Driver Ratios	0.5	0.6	0.7	0.8	0.9	1.0	1.1	1.2	1.3	1.4	1.5	1.6	1.7	1.8	1.9	2.0
Ratio of Arithmetic to Geometric Mean	1.061	1.033	1.016	1.006	1.001	1.000	1.001	1.004	1.009	1.014	1.021	1.028	1.035	1.043	1.052	1.061

The choice is ours.

Going back to the original discussion on the two extremes of the Product or the Geometric Mean of the Driver Changes, we may have noticed that these extremes lie approximately at either end of the likely values of the power values highlighted by Chilton (see Chapter 2) in his 'Power Rule', which suggests that power values are normally in the range of 0.35 to 0.85. In this case the Geometric Mean suggests a Power Rule based on 0.33, and full driver independence is indicated by the Power Rule of 1, i.e. the simple product of the drivers.

For the Formula-philes: Geometric Mean ≈ Arithmetic Mean for similar Factors

Consider two estimating drivers D_1 and D_2 which are to be used to develop an estimate by analogy for a new product by comparing it with an existing product. Suppose the values of the two drivers for the existing product are $D_{1,0}$ and $D_{2,0}$. The corresponding drivers for the new product are determined as $D_{1,1}$ and $D_{2,1}$.

The Factors $\boldsymbol{F_1}$ and $\boldsymbol{F_2}$ which reflect the differences in the two drivers can be defined as:

$$F_1 = D_{1,1} / D_{1,0}$$
$$F_2 = D_{2,1} / D_{2,0} \qquad (1)$$

As $\boldsymbol{F_1}$ and $\boldsymbol{F_2}$ are constants for the case in question, one can be expressed as a multiple of the other:

$$F_2 = \gamma \, F_1 \qquad (2)$$

The Arithmetic Mean of the Factors, A-Mean, is:

$$\text{A-Mean} = \frac{F_1 + F_2}{2} \qquad (3)$$

The Geometric Mean of the Factors, G-Mean, is:

$$\text{G-Mean} = \sqrt{F_1 F_2} \qquad (4)$$

From (3) and (4) the ratio of A-Mean to G-Mean is:

$$\frac{A - Mean}{G - Mean} = \frac{F_1 + F_2}{2\sqrt{F_1 F_2}} \qquad (5)$$

Substituting (2) in (5):

$$\frac{A - Mean}{G - Mean} = \frac{F_1 + \gamma F_1}{2\sqrt{F_1 \gamma F_1}} \qquad (6)$$

Simplifying (6) to remove $\boldsymbol{F_1}$ from both the numerator and denominator:

$$\frac{A - Mean}{G - Mean} = \frac{1 + \gamma}{2\sqrt{\gamma}} \qquad (7)$$

Consider the case where one factor is a third larger than the other ($\gamma = 4/3$), the ratio of the two Means (7) takes the value:

$$\frac{A - Mean}{G - Mean} = \frac{7}{12}\sqrt{3}$$
$$= 1.0104$$

... which is around 1% difference

Before we get too excited by all this (*oh, that was just me, was it?*), we need to recognise that the Geometric Mean here was based on three drivers, hence the 'one third power rule'. If we had only had two drivers, the Geometric Mean would be equivalent to a 'one half power rule' or square root rule – still in the range of Chilton's observations. However, if we had had four Drivers the Geometric Mean would indicate a 'one quarter power rule' which lies outside the range indicated by Chilton. We can postulate that this might suggest that we are drifting away from purely Primary Drivers and are including Secondary Drivers. The more Drivers we consider in a basically simple technique, the more likely we are to have interaction and duplication between them.

Caveat augur

If we believe that there are three potential Drivers, but in a particular scenario one of the Drivers has the same value as that for the reference case, we should not ignore it as the Geometric Mean for two Drivers will be different to the Geometric Mean for the three Drivers.

For instance, if we were to consider an option in the example from Table 5.13 for the new ship that requires the Maximum Speed to be only 20 kt (the same as our cost reference item), then we might be tempted to exclude it from our analogical relationship as a superfluous term. **Do not give in to this temptation!** As we can see from Table 5.15, a Factor based on two overlapping Drivers would give us a significantly different estimate than one based on three overlapping Drivers. In essence we have to make a decision; either we believe that Maximum Speed is a genuine cost Driver or it is not. Just because the Driver has the same value as our cost reference point, does not mean we can

Table 5.15 Impact of Ignoring a Valid Driver with a Factor of 1

Cost Driver	Reference Ship	New Ship	Cost Driver Factor	Cost Driver Factor
Gross Tonnage (gt)	3200 gt	3600 gt	1.125	1.125
Personnel on Board	21	24	1.143	1.143
Maximum Speed	20 kt	20 kt	1.000	Ignored

Product of Factors	1.286	1.286
Geometric Mean of Factors	1.087	1.134
	Based on 3 Factors	Based on 2 Factors

ignore it; to do so may be the difference between winning, or losing a competitive bid (*or at least save us from the embarrassment of getting it wrong!*).

5.6 Sensitivity Analysis on Factors, Rates and Ratios

Unless we are just using Factors, Rates and Ratios to convert rule-based normalisation adjustments, we should consider how sensitive our estimates will be based on small or large changes in the values we use. It is unlikely that the Factors, Rates and Ratios we select will be precisely correct, so it is good practice to test the impact of differences in the driver values on the estimate and so create a range estimate.

The question we need to resolve is to what extent should we 'flex' our selected values? There is always the tried and trusted method of 'common sense'. (*However, as my very astute wife likes to point out, I have no common sense, so I am hardly qualified to talk about it!*)

Another approach to performing a Sensitivity Analysis favoured by many is that of using a **NAFF** approach, which is short for '*plus or minus a Nice Arbitrary Fixed Factor*' where the Arbitrary Fixed Factor is a nice rounded fixed percentage chosen at random, e.g. ±5%, ±10%. That is not to say that the values are not appropriate, but the manner in which they are chosen is often random rather than by a reasoned and rational approach.

Also, the addition or subtraction of a fixed percentage implies that we believe that the Metric is symmetrically distributed, which often will not be the case as we discuss in Volume II Chapter 3. Where we have data, we can usually do better than use a NAFF approach...

5.6.1 Choosing a Sensitivity Range quantitatively

Ideally, if we have taken the trouble to select an appropriate value that is one of the three principal Measures of Central Tendency (Volume II Chapter 2), we should also consider the scatter of data around the chosen value. In Volume II Chapter 3 on Measures of Dispersion and Shape, we will discuss a number of measures which could be useful:

Measures of Dispersion:

- Minimum and Maximum
- Mean or Average Absolute Deviation
- Median Absolute Deviation (*otherwise known as the MAD measure*)
- Standard Deviation

Measures of Shape:

- Skewness
- Kurtosis or Peakedness

Measures of Dispersion and Shape:

- Confidence Levels, Limits and Intervals
- Quantiles and Interquantile Ranges

The latter group is probably the most useful to us if the data is available. Here we can make a definitive choice of values which we can express as rounded extrinsic percentage factors if we prefer – but they are unlikely to be in nice 5% increments in many cases. Regardless of whether the data is symmetrical or skewed, 'peaky' or 'flattish', we can use various Interquantile Ranges to approximate the level of sensitivity we choose:

- 25% to 75 % Confidence Limits   => Interquartile Range   *(middle 50%)*
- 20% to 80 % Confidence Limits   => Interquintile Range   *(middle 60%)*
- 10% to 90 % Confidence Limits   => Interdecile Range   *(middle 80%)*
- 5% to 95 % Confidence Limits   => Intervigintile Range   *(middle 90%)*
- 2.5% to 97.5% Confidence Limits => Interquadragintile Range   *(middle 95%)*

Figure 5.7 and Table 5.16 illustrates the use of an Interquintile Range to choose sensitivity limits objectively for the example we discussed in Section 5.3 (Table 5.3).

Quintiles and other Quantiles are discussed more fully in Volume II Chapter 3. To calculate the Quintiles in Figure 5.7 we have used the closed **PERCENTILE. INC** function in Microsoft Excel rather than the open **PERCENTILE.EXC**

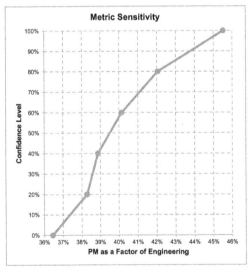

Figure 5.7 Choosing a Sensitivity Range Based on an Interquintile Range

Table 5.16 Choosing a Sensitivity Range Based on an Interquintile Range

Estimated Engineering Hours	Quintile Based Sensitivity Range	PM Confidence Level	PM to Eng %	Estimated Project Management Hours
5,320	Optimistic >	20%	38.3%	2,036
	Average >	59%	40.0%	2,128
	Pessimistic >	80%	42.1%	2,237

version. For reasons, which we will discuss in Volume II Chapter 3, the latter would be a better function to use instead of the former, but we will not repeat the explanation here.

Now, we may make the observation that ±5% of the chosen Factor of 40% (giving 38% to 42%) is not a million miles away the Interquintile Range in this example. That is true, and it is down to personal preference whether we go with the 'imprecise' calculation above (*let's not kid ourselves that it is truly precise – it is based on the sample available and a different sample would give a different result*), or the rounded view of 38% to 40%. However, we now have evidence that to support the ±5% and can even say that the Confidence Levels for a Factor of 38% and 42% are 16% and 79% respectively.

If we want to express the 'best and worst case' scenarios, we can use the Minimum and Maximum, or in the case of something we believe to be Normally Distributed, three Standard Deviations either side of the Mean (see Volume II Chapter 4).

5.6.2 Choosing a Sensitivity Range around a measure of Central Tendency

In general, when we are using a number of Factors, Rates or Ratios in a multiplicative model (unless the values are highly correlated and one can be largely substituted by another), we do not need to take the worst case scenario to test the sensitivity of the end result. This is due to the compounding effect of low probabilities; not all the good things in life happen together just as not all bad things in life happen together either (*it just seems that way – unless the distribution of good and bad is right skewed, but that is getting us into the realms of philosophy and out of scope of this compendium!*).

For instance:

Probability of getting values for variable 1 less than its Optimistic Value = 10%
Probability of getting values for variable 2 less than its Optimistic Value = 10%
Probability of getting values for both variables less than their Optimistic Values = 1%

The mirror image of this argument can be made for the Pessimistic end of the range For a less conservative approach to choosing a sensitivity range than one based on the best or worst case scenarios, we might want to consider a range based on one Standard Deviation on either side of the Mean as illustrated in Figure 5.8. We will cover the theory of this in Volume II Chapter 3.

We should not read too much into the Confidence Interval in this example as it will depend on the degree of dispersion and peakedness of the data. If we were to look at a Normal Distribution, as we have already discussed briefly, one Standard Deviation on

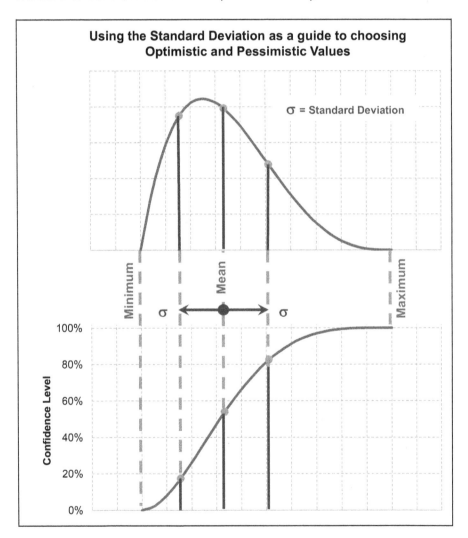

Figure 5.8 Using the Standard Deviation as a Guide to Choosing Optimistic and Pessimistic Values

either side of the Mean will give us a 68% Confidence Interval, whereas two Standard Deviations either side will give 95% Confidence Interval. In this particular example of a Right Skewed Beta Distribution, if we were to deduct two Standard Deviations from the Mean, we would be lower than the Minimum! (*A clear oxymoron, if there ever was one.*)

For an even less conservative approach, i.e. one that still considers a reasonable range to test sensitivity, but slightly narrower still than one based on one Standard Deviation, we might want to consider the following 'rule of thumb' guide based on the two Measures of Absolute Deviation, which we will be discussing in Volume II Chapter 3. Table 5.17 proposes a combination of both Mean and Median Absolute Deviations to establish sensitivity ranges, depending on the degree of skewness and peakedness.

Note: whilst this approach is quite valid, we may want to reflect on whether it really gives any tangible benefit over using the Standard Deviation.

Figure 5.9 illustrates the use of the two Absolute Deviation measures in the case of a Positive or Right Skewed Beta Distribution. The Median Absolute Deviation (MAD) is used to the left of the Median, whereas the Average Absolute Deviation (AAD) is used to the right of the Mean.

The mirror image of this would be used on a Negatively or Left Skewed distribution as the Mean and Median would flip positions (see Volume II Chapter 2).

Table 5.17 Choosing Sensitivity Ranges Based on Measures of Absolute Deviation

		Skewness		
		Left or Negatively Skewed	*Symmetrical*	*Right or Positively Skewed*
Kurtosis or Peakedness	**Positive Excess Kurtosis or Peakier than Normal**	**Optimistic =** Mean minus Average Absolute Deviation	**Optimistic =** Mean (or Median) minus Average (or Median) Absolute Deviation	**Optimistic =** Median minus Median Absolute Deviation
	Zero Excess Kurtosis or Normal	**Pessimistic =** Median plus Median Absolute Deviation	**Pessimistic =** Mean (or Median) plus Average (or Median) Absolute Deviation	**Pessimistic =** Mean plus Average Absolute Deviation
	Negative Excess Kurtosis or Flatter than Normal	**Optimistic** = Minimum **Pessimistic** = Maximum		

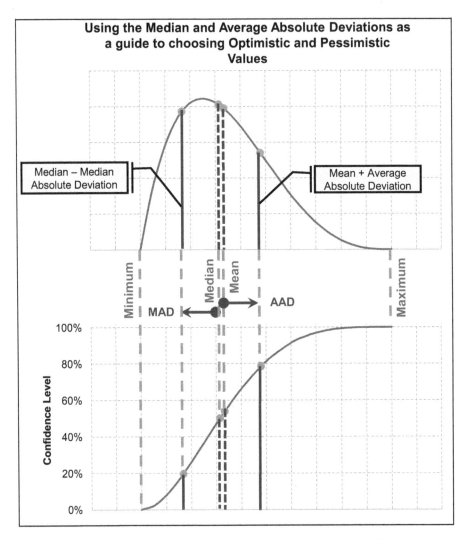

Figure 5.9 Using the Median and Average Absolute Deviations as a Guide to Choosing Optimistic and Pessimistic Values

In the case of a symmetrical distribution, the Mean = Median, and MAD = AAD, hence there is no need for the switching.

Note: From a pragmatic perspective, these deviation-based approaches are probably more appropriate where sensitivity is going to be expressed digitally, in terms of a reasonable level of optimism or pessimism, rather than a best case and worst case scenario. If the values are to be used as inputs to a Monte Carlo Analysis, the Minimum/

Maximum spread is considered to be more appropriate (refer to the discussion in Volume V Chapter 3).

5.6.3 The triangulation option

If we have chosen the Mode (or what we believe to be the Mode) as our Metric value, and we have a view of the Minimum and Maximum value, but we do not know the underlying distribution in order to use a Confidence Interval or Interquantile Range, then, rather than use the best and worst case scenario, we could take optimistic and pessimistic sensitivity bounds on based on:

Lower Bound: The mid-point between the Minimum and Mode
Upper Bound: The mid-point between the Mode and the Maximum

We can take this approach if we make the assumption that the distribution can be approximated by the Triangular Distribution formed by the Minimum, Modal Height and Maximum (see Volume II Chapter 4). This approach will give us an asymmetric 75% Confidence Interval (see Key Properties of a Triangular Distribution in Volume II Chapter 4), for which we can calculate the implied Confidence Limits quite easily should we wish (see Table 5.18).

5.6.4 Choosing a Sensitivity Range around a High-end or Low-end Metric

There will be occasions when we have elected to choose a high- or low-end value for a Metric in preference to one of the Measures of Central Tendency. In such cases we might want to consider the following sensitivity ranges that maintain the consistency of the rationale for choosing the high- or low-end Metric:

Table 5.18 Creating a Simple 75% Asymmetric Confidence Interval for a Metric

Metric Value		Cumulative Probability	Comment
Minimum	10 hrs	0%	
LH Range Mid-point	11 hrs	6.25%	< 25% of Left Hand Range
Mode	12 hrs	25%	< Left Hand Range
RH Range Mid-point	15 hrs	81.25%	< 100% - 25% of Right Hand Range
Maximum	18 hrs	100%	< Overall Range
Left Hand Range	2 hrs	25%	< Mode - Minimum
Right Hand Range	6 hrs	75%	< Maximum - Mode
Overall Range	8 hrs	100%	< Maximum - Minimum

High-end Metric
Optimistic: Mean, Mode or Median
Pessimistic: Maximum

Low-end Metric
Optimistic: Minimum
Pessimistic: Mean, Mode or Median

The choice of Mean, Mode or Median in the above should be influenced by the rationale behind choosing the basic Metric, i.e. was it chosen in relation to being higher or lower than the Mean, Mode or Median?

5.6.5 Choosing a Sensitivity Range when all else fails

If we don't have any information on the distribution or any Measures of Dispersion or any view of the basic shape, then there is always the NAFF (Nice Arbitrary Fixed Factor) approach raised earlier, but before we resort to that we should really be asking ourselves:

- How much better might this be?
- Given the answer to the above, how much worse might it be, bearing in mind that it is often the case that reality is skewed one way or another? It is much easier to under-perform than over-perform.

We might also want to consider plotting Sensitivity Charts depicting the incremental percentage variation in the value chosen, as we did in Tables 5.9 to 5.11 in Section 5.5.2.

Let's review the sensitivity of the analogy we made in the first of these in relation to changes in our assumption of the proportion of Total Cost attributable to Labour Costs. We will see from Figure 5.10, based on the data in Table 5.19, that the weighted average adjustment which forms the basis of the Factor to be used in the analogy varies very little ... even if we extend beyond the range that we feel is appropriate. This is because the changes in the two Cost Drivers are of a similar order of magnitude. This should give us a nice soothing feeling that our analogy is relatively stable (... *a bit like an icepack is to a headache to use another analogy*).

Note: The Expected Range shown here relates to these examples only, and it should not necessarily be considered as an appropriate range in general circumstances.

On the other hand, if we consider the same problem where the Cost Driver changes are not of the same order of magnitude, in fact in Table 5.10, they change in opposing directions, giving us the result in Figure 5.11 (based on Table 5.20). This gives us a far less comfortable feeling.

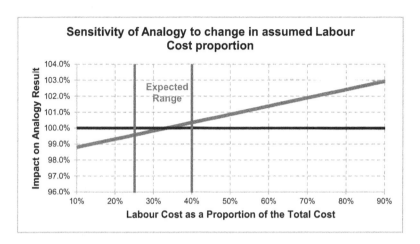

Figure 5.10 Sensitivity of Analogy to Change in Assumed Labour Cost Proportion – Example 1

Table 5.19 Sensitivity of Analogy Result to a Change in Labour Cost Proportion – Example 1

	Number of Items	Weight (kg)
Product A	50	25
Product B	60	28.5
Ratio	120%	114%

	Product A Input Weighting		Weighted Average Change Factor	Sensitivity Relative to Base Assumption
	Labour Costs	Non Labour Costs		
	10%	90%	1.146	98.8%
	20%	80%	1.152	99.3%
	25%	75%	1.155	99.6%
Expected Range	30%	70%	1.158	99.8%
	33%	67%	1.160	100.0%
	40%	60%	1.164	100.3%
	50%	50%	1.170	100.9%
	60%	40%	1.176	101.4%
	67%	33%	1.180	101.7%
	70%	30%	1.182	101.9%
	75%	25%	1.185	102.2%
	80%	20%	1.188	102.4%
	90%	10%	1.194	102.9%

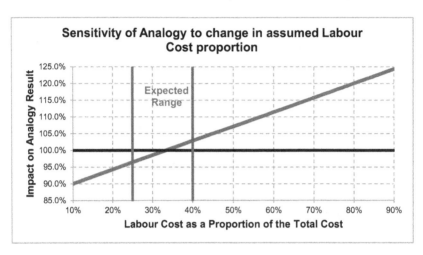

Figure 5.11 Sensitivity of Analogy to Change in Assumed Labour Cost Proportion – Example 2

Table 5.20 Sensitivity of Analogy Result to a Change in Labour Cost Proportion – Example 2

	Number of Items	Weight (kg)
Product A	50	25
Product B	60	20
Ratio	120%	80%

	Product A Input Weighting		Weighted Average Change Factor	Sensitivity Relative to Base Assumption
	Labour Costs	Non Labour Costs		
	10%	90%	0.840	90.0%
	20%	80%	0.880	94.3%
	25%	75%	0.900	96.4%
Expected Range	30%	70%	0.920	98.6%
	33%	67%	0.933	100.0%
	40%	60%	0.960	102.9%
	50%	50%	1.000	107.1%
	60%	40%	1.040	111.4%
	67%	33%	1.067	114.3%
	70%	30%	1.080	115.7%
	75%	25%	1.100	117.9%
	80%	20%	1.120	120.0%
	90%	10%	1.160	124.3%

In this case the range is wider. In relation to the 'expected range' for the proportion of the Total Costs attributable to Labour Costs, the sensitivity is not too great (-3.6%, +2.9%), but as we step outside of this 'comfort ground' the percentage sensitivity becomes more of a concern.

Caveat augur

This sensitivity of an estimate to variation in an assumption or Driver value is **not** a measure of the accuracy of an estimate; it is more of a measure of precision, and should be regarded as a measure of robustness of the estimate in relation to changes or errors in assumptions.

5.7 Chapter review

In this chapter we have discussed the differences and similarities between three different Estimating Metrics: Factors, Rates and Ratios and where we might use them. In essence, all three are derivatives of the same basic principle that underpins many estimating methods and techniques.

We will use Factors, Rates and Ratios in Normalising Data (Chapter 6) to allow the data to be used equitably in comparison with something else. Their use is fundamental to the creating estimates by Analogical Methods.

Rate Metrics are expressions of how something changes in relation to some measurable driver or parameter, and would be expressed in the form of a [Value] per [Unit], such as Cost per Hour, Hours per kilogram etc. They are used to convert one estimating entity into another and in so doing convert the units of measurement, e.g. an estimate of hours is converted into an estimate of cost.

Factor Metrics are used to express one value as a percentage of another value. By implication, as percentages, Factor Metrics are dimensionless and do not change the unit of measurement, and merely scale values up or down in relation to another. Factor Metrics can occur in two sorts: Intrinsic and Extrinsic. Extrinsic Factors express the delta difference between two values, e.g. there is 10% difference, whereas Intrinsic Factors are inclusive of the value to which the comparison is made, e.g. one product is 110% of the size of another.

Ratio Metrics are very similar to Intrinsic Factor Metrics in that they express a scale relationship between two values that are measured in the same units.

In using multiple Factors, Rates and Ratios, we discussed the need to understand how the various Primary, and to a lesser extent Secondary Drivers, interact, and what options we have in different circumstances:

- Is it an additive or multiplicative relationship?
- Are any of the Drivers independent of each other, or are they partially or fully correlated?

As estimating is never an exact science (*implicit within the name*), it is recommended that we test the Sensitivity of the estimates produced in relation to variations in the Factors, Rates and Ratios we choose. The primary techniques we can use to scale the range of this Sensitivity Analysis are based on the Measures of Dispersion and Shape that we will discuss in more detail in Volume II Chapter 3. (*So that is why that chapter is there – and if you have had a sneaky peek you may have been thinking that I was just being self-indulgent or sadistic?*)

References

Augustine, NR (1997) *Augustine's Laws (6th Edition)*, Reston, American Institute of Aeronautics and Astronautics, Inc.

Department for Transport – Driving Standards Agency (2015) *The Official Highway Code*, London, The Stationary Office.

Lang, HJ (1947) 'Cost relationships in preliminary cost estimation', *Chemical Engineering*, Oct.

Lang, HJ (1948) 'Simplified approach to preliminary cost estimates', *Chemical Engineering*, June.

ICEAA (2009) *Cost Estimating Body of Knowledge*, Vienna, International Cost Estimating and Analysis Association.

Turré, G (2006) 'Plant capacity and load', in Foussier, P, *Product Description to Cost: A Practical Approach, Volume 1: The Parametric Approach*, London, Springer-Verlag pp. 141–143.

6 Data normalisation – Levelling the playing field

'When was anything ever 'normal' in the life of an estimator?' I hear you muse.

The key to creating sound and robust estimates is that we base our analysis and judgement on facts that represent the past and the present, and assumptions that represent the future (*perhaps an oversimplification, but true in essence*). For this, we require data and knowledge that express factual information in a meaningful manner. When we are collecting data for the purposes of estimating, we will usually collect anything that appears remotely useful that we can recycle. Laudable though this eco-friendly approach may seem, it does present us with the problem that the data in its raw form often cannot (or, rather, should not) be used. To use it in its raw form may be considered reckless and may lead to poor management decisions being made.

So, what do we do? Dispense with the facts if they are so problematic, and take a wild guess instead? Certainly not! Instead, we need to understand the context that existed around those facts and gather those as well. Data normalisation is the process of making reasonable adjustments to the data (i.e. the facts) in order to make fair and equitable comparisons with other data under the same or similar conditions. (*Don't read this as a licence to fictionalise or fiddle the facts – a repeatable procedure and audit trail is essential.*) Remember that TRACEability is key to good professional estimating practice.

One authoritative source defines the verb 'to normalise' (in the mathematical sense) as:

> In some sectors of industry, the process of normalisation may be referred to as '*benchmarking*', implying that the elements to be compared are being measured from the same benchmark. This use of the term is compatible with the act of normalisation, but unfortunately, the term '*benchmarking*' can also imply something completely different, more synonymous with '*best in class*' or '*best practice*'.

1. '*Bring to a normal or standard state*
2. *Mathematics: multiply by a factor that makes the norm equal to a desired value (usually 1)*'

Stevenson, Angus & Waite, Maurice (Eds) (2011) *Concise Oxford English Dictionary (12th Edition)*, definition of Normalise
By Permission of Oxford University Press

However, as we will see shortly, factoring is only one means of normalising data for estimating. Addition, subtraction, segregation (including exclusion) are all other valid techniques we can use, aimed at ensuring that we have a '*level playing field*' when we make comparisons with data for estimating. We'll be using the more general definition here.

It is important that we realise that anything we can '*normalise*' can also be '*unnormalised*' again – it is a reversible process, so long as we document what we have done.

So, with that in mind, how might we define Data Normalisation then …?

Definition 6.1 Data Normalisation

Data Normalisation is the act of making adjustments to, or categorisations of, data to achieve a state where data the can be used for comparative purposes in estimating.

A word (or two) from the wise?

'It's important that athletes can compete on a level playing field …'
Paula Radcliffe
British Athlete
Courtesy of Guardian News & Media Ltd

We would probably all agree with the extract from an interview with Paula Radcliffe in which she is reported (Bull, 2012) to have commented on the importance of athletes competing 'on a level playing field'. This may seem to be somewhat of an obvious statement to make, a bit of a 'no-brainer'. However, she wasn't using it in the literal sense. (*She was referring to the need to eliminate cheating in the sport to encourage youngsters to take it up.*) In our context, the level playing field is a reflection of the need to allow each data point to 'compete' on an equal footing as all other data points.

However, it should be equally obvious in relation to estimating that if we don't '*level the playing field*' in terms of the historical or current data and the context in which it exists, and that for some future state and value that we are trying to estimate, then we run the risk of being uncompetitive (too excessive, too late, too low a performance etc.) or of being naively aggressive (too bullish, too early, too high a specification etc.).

6.1 Classification of data sources – Primary, Secondary and Tertiary Data

Before we use any data for estimating, we really should understand its pedigree. Data for estimating can emanate from a number of different sources, too many to mention here other than in the generic sense, which we will classify as Primary, Secondary or Tertiary Data. We should consider this segregation into these three categories as a guide only rather than as an absolute division, and its inclusion here is more to guide our thinking.

6.1.1 Primary Data

Definition 6.2   Primary Data

Primary Data is that which has been taken directly from its source, either directly or indirectly, without any adjustment to its values or context.

We would think that data extracted from a primary source would be the most appropriate data to use in estimating, even if sometimes it is not the most convenient. Sometimes, it is, but:

- Primary Data is not always as 'clean' as we would like it to be, especially where the data input is susceptible to human error or indiscipline in working practices. If this data has already been scrutinised and the errors removed or adjusted in an appropriate manner, then the resultant dataset may be more convenient for us to use. However, it is no longer Primary Data by our definition.
- In that case, we can question whether it is always appropriate to use data direct from source, especially if the data contains known or suspected errors. If the estimate we create eventually manifests itself in the future in the pool of actual data on which it had been based, then the answer must surely be that it is appropriate, especially if those type of errors are systemic.

We should not be tempted to make adjustments for suspected but unconfirmed errors in Primary Data. To do so runs the risk of introducing statistical bias into our estimates. The main benefit to estimators of using Primary Data is that its source and credibility (*warts and all*) are more easily understood.

In the case of an 'Extrapolation from Actuals' to create an 'Estimate To Completion' or ETC (for example), or for an estimate for some follow-on work, we may well be using Primary Data … unless we know that something fundamentally atypical has occurred.

Table 6.1 Example of Primary Data

Engine bay door manufacturing and assembly time		
Year	Recorded Hours	Comment
2008	1,437	Hours exclude Inspection (indirect activity)
2009	3,443	Hours include Inspection as a direct charge
2010	4,321	Hours include Inspection as a direct charge
2011	3,567	Hours include Inspection as a direct charge
Total	12,768	Total Direct Hours chargeable to Contract

In the example in Table 6.1, there has been a change in the organisation's accounting conventions in terms of the type of activity that can be charged direct to contract (Direct Charge) as opposed to one that must be recovered through the labour charging rate (Indirect Charge). In this case, during the manufacture and assembly of an Engine Bay Door, the activity of inspection was changed from an Indirect to Direct Charge at the end of 2008. The Recorded Hours by year (unadjusted) constitute Primary Data. In this case we need to normalise the data and in so doing, this creates Secondary Data.

6.1.2 Secondary Data

There will be instances as estimators where we will find it more convenient to use data that is not strictly speaking Primary Data as we have defined it here, but has undergone some controlled change to present it in a more useable or acceptable form. This would be Secondary Data.

Definition 6.3 Secondary Data

Secondary Data is that which has been taken from a known source, but has been subjected to some form of adjustment to its values or context, the general nature of which is known and has been considered to be appropriate.

If there has been a one-off correction to Primary Data to remove an obvious error, then we are not going to start splitting hairs and demand that we call that Secondary Data, especially where that data change has been recorded in the source system as a transaction. However, where we have made adjustments outside of the source systems to take account of such changes, or we have transformed the data more fundamentally to take account of differences in the context that we wish to use the data, then it becomes Secondary Data. The key message is that Secondary Data is derived from Primary Data and there is a clear audit trail of the changes made.

We may find that Secondary Data is more convenient to use than Primary Data where we are creating an estimate with the same data used for another estimate; the rationale being, we don't have to repeat the same exercise to get to the same start point? If we have insufficient time (*sound familiar?*) to prepare an estimate from Primary Data, then we may still find it beneficial to take Secondary Data that has been adjusted to a defined and recorded state, and that the conclusion drawn is relevant or bridgeable to the case in question.

In developing Secondary Data we may make a number of adjustments due to:

- Differences in work content
- Effects of inflation or escalation
- Changes in accounting policies (direct versus indirect hours or costs)
- Alternative measurement systems or scales (e.g. °F to °C, inches to centimetres etc.)

Table 6.2 builds on our previous example of Primary Data (in Table 6.1) by adjusting the Recorded Hours for 2008 by an amount that we feel is a reasonable estimate of the Inspection Hours that would have occurred had they been a Direct Charge. This adjusted data then constitutes Secondary Data. The important thing is to maintain a record in the Basis of Estimate of the nature, reason and justification for the adjustment: TRACEability is key.

6.1.3 Tertiary Data

Definition 6.4 Tertiary Data

Tertiary Data is data of unknown provenance. The specific source of data and its context is unknown, and it is likely that one or more adjustments of an unknown nature have been made, in order to make it suitable for public distribution.

Table 6.2 Example of Secondary Data

Year	Total Recorded Hours	Comment	Inspection Hours Included	% Inspection Hours	Estimated Inspection Hours	Normalised (Adjusted) Hours
2008	1,437	Hours exclude Inspection (indirect activity)	#N/A		151	1,588
2009	3,443	Hours include Inspection as a direct charge >	323	9.4%		
2010	4,321	Hours include Inspection as a direct charge >	415	9.6%		
2011	3,567	Hours include Inspection as a direct charge >	338	9.5%		
Total 09-11	11,331	Total Direct Hours chargeable to Contract	1,076	9.5%		

Engine bay door manufacturing and assembly time normalised for accounting change in 2009

Now, you might well be thinking '*why on earth would I use data of unknown provenance?*' Well, we may not have a choice – it may be better than a wild guess! Consider the case of an innovative design, or the impact of a new technology. It is quite possible that we may not have any data of our own, but there is a greater chance that someone somewhere sometime has done something, somehow, that can be used as a comparator (...*yes, that just goes to show that there are a lot of 'somes' involved in estimating!*). We probably wouldn't use the data as the basis of a contractual agreement without further research and validation, but it is something we can use to start the proverbial ball rolling on our level playing field!

Typically, the source of Tertiary Data may be publications on the internet, in trade or professional journals, academic publications or information from suppliers and vendors (e.g. Catalogue Prices), or the database that underpins Commercial Off-the-Shelf Parametric toolsets. Its source may be known only in a generalised sense, but the composition of the data and the nature of any adjustments made will be unknown to us.

In short, Tertiary Data is that which has been 'normalised' or 'modified' by a third party or not as may be the case, to make it suitable for public consumption.

There is an argument that Tertiary Data is less likely to be adjusted further as we are unlikely to know either the adjustments made already in making the data public, or the size of any adjustment required to take account of relevant parameters that reflect the context of the Tertiary Data, e.g. size data.

Typically, we might consider using Tertiary Data when:

- There are no internal sources of data available
- The data is being used for Market Analysis, or general 'sizing' exercises (Rough Order of Magnitude or RoM)
- As a check value for our internal estimate

There is a strong case that if the audit trail for internal Secondary Data has been 'lost', then to all intents and purposes it has become internal Tertiary Data of unknown provenance.

In the example in Table 6.3, the data from Company A and Company B is Tertiary Data as we know nothing about the veracity of the data or its context, and must take it 'at face value' or reject it. Even though we have some explanatory notes for Company A, the researcher has not clarified to us whether the time expended by operators in performing self-inspection activities is included in the numerator and/or the denominator.

Table 6.3 Example of Tertiary Data

Benchmarking exercise on inspection activities		
Organisation	Inspection % Delta to Production	Comments
Company A	8%	Company employs certified self-inspection scheme to supplement Inspection
Company B	10%	No additional information provided
Own Company	9% to 10%	Company uses dedicated inspectors only

6.1.4 Quarantined Data

We might consider that factoring up actual Work In Progress extracted from a Primary Source to get an 'Estimate At Completion' (EAC) is another example of creating Secondary Data by normalising for the percentage completion to 100%. However, it is questionable science. How often does that last 10% take or cost proportionately more than the work in progress so far? This is another fine example of Augustine's Law of Unmitigated Optimism (Augustine, 1997)!

The less we have completed, the stronger the case is that we only have something that is better construed as Tertiary Data. The difference between what was predicted and the final outturn is unknown at this stage, hence the EAC is of unknown provenance.

Instead we might want to consider quarantining the data, so that we neither use it nor lose it. We should consider it to be better practice to compile an estimate based on known actual data (in a fully complete sense) and to use the EAC based on factored Work In Progress simply as a check value to validate that the estimate is pitched at the right level.

6.2 Types of normalisation Methods and Techniques

The Methods used for Data Normalisation, which essentially is creating estimates of an alternative reality, are the same Estimating Methods outlined in Chapter 2, comprising Analogous, Parametric and 'Trusted Source' adjustments.

There are relatively few basic Data Normalisation techniques. These can be summarised as:

a)   Element Addition or Subtraction
b)   Factors Rates and Ratios, or Analogical
c)   Formulaic or Parametric
d)   Segregation or Separation

We can use Factors, Rates and Ratios just as we did in the last chapter in order to create an estimate by analogy. In essence, that is what we are doing when we normalise values in this manner; we are using Factors, Rates or Ratios to change a past or present actual value into one that can be used as an Estimate Reference point. (*This sounds like an Analogy of an Analogy.*) Now some of us may well be thinking if we are using Factors, Rates and Ratios to normalise our data in order to create an estimate by analogy, we are in effect just using multiple Factors, Rates and Ratios as we did in Chapter 5. Does this mean that we could just roll this chapter and the last chapter together? Well, to a point, yes, but in general, no, (*there I go sitting on that fence again*) because ...

- There are other normalisation techniques that do not use Analogical Factors, Rates or Ratios
- It is sometimes helpful to use Factors, Rates and Ratios to normalise data where there is an objective means of doing so, and leave the more subjective adjustments required to the Analogical Estimate creation step. In other words, we might apply two factors, rates or ratios: one for known quantified changes and one for more subjective or qualitative changes

---

Quantitative and Qualitative Adjustments

- **Quantitative Adjustments**: those for which we can make an adjustment in an objective manner using a fixed or variable rule. These will include adjustments for differences in units of measurement, cost or value.
- **Qualitative Adjustments**: those for which we make adjustments in a more subjective or judgemental manner. These will include adjustments for differences in assumed relative complexity, or general size categorisation.

All the adjustments are inherently Quantitative but the value used could be based on a Qualitative decision, e.g. 10% more complex.

We can use parametric modelling to identify appropriate adjustments to make; these can be in relation to single or multiple estimate drivers. These adjustments could be expressed in terms of Factors, Rates or Ratios, but the key difference lies in that the analogical technique uses pre-determined or assumed values, whilst a Parametric Technique looks at the 'best fit', or at least a predetermined formulaic technique. The latter also allows a means of normalising or 'un-normalising' the future values.

6.3 Normalisation can be a multi-dimensional problem

There are many different 'dimensions' that we may need to consider in our normalisation step, and these might lead to different approaches to normalising of our data. We may have differences due to known or perceived errors, scaling, volume or quantity effects, date related (absolute or relative), role or purpose, work content or scope, complexity or difficultness etc. In some cases, the adjustments we need to make are rule based, whereas others are more to do with categorisation.

In reality, the greater the difference between the 'thing' we are trying to estimate and the 'thing' or 'things' for which we have data, then the greater the number of dimensions we will have to consider normalising, and/or the greater the normalisation adjustments are likely to be and the more complex our Data Normalisation process becomes. We may decide that there are some instances where a data item is considered to be too different or inappropriate to compare – the archetypal *apples and oranges*; in such a case we will probably elect to pursue the somewhat terminal normalisation technique of 'exclusion', i.e. rejecting the data.

As the Data Normalisation step is an important element of the Basis of Estimate, all the normalisation adjustments should be TRACEable (Transparent, Repeatable, Appropriate, Credible and Experientially-based) as highlighted in Chapter 3. In many cases, the limiting factor on our ability to normalise data will be how well the supporting contextual data has been recorded in the past.

6.3.1 Error related

If we know of, or have a strong case to support our perception of there being an error in actual data recorded, then the first step in normalising our data is to make those adjustments before all others – and to record what we have done and why (for the next person who accesses the data, and to mitigate our failing memories – see Table 6.4).

We should not be tempted to make adjustments to any data that is an outlier simply to make it fit the data pattern we want; we need to be sure (*balance of probabilities and all that*) that there is a genuine error to be corrected otherwise we will be failing our TRACE mantra.

6.3.2 Volume, quantity or throughput related – Economies of Scale

There may be some adjustments that we need to make as a result of a difference between the planned quantity to be produced and those we have at our disposal in terms of historical values. Collectively, we might want to refer to these as 'Economies

Table 6.4 Example of Data Normalisation to Correct Known Errors

Work Breakdown Structure Element	Budget	Invoices Paid	Cost Transfer	Corrected Actual Costs	Comments
1.1.1	$2,145	$2,131		$2,131	
1.1.2	$2,500	$2,578		$2,578	
1.1.3	$585	$1,588	-$996	$592	Transferred to 1.3.1 – Purchase Order input error
1.2.1	$5,300	$4,900		$4,900	
1.2.2	$3,255	$3,255		$3,255	
1.2.3	$0	$0		$0	
1.2.4	$0	$0		$0	
1.3.1	$1,877	$1,002	$996	$1,998	Transferred from 1.1.3 – Purchase Order input error
1.3.2	$238	$240		$240	
2.1.1	$1,000	$990		$990	
2.1.2	$1,000	$0	$975	$975	Previously charged incorrectly to Sustainment Contract
Total	$17,900	$16,684	$975	$17,659	

of Scale'. These adjustments may simply be to determine a '*per unit*' value but we may have to consider the impact of the batch or lot size in relation to any cost, time or duration that occurs once for each lot rather for each unit, giving us the terms '*Set*' and '*Run*':

- 'Set' values relate to the one-off values that occur per batch or per lot
- 'Run' values relate to the recurring values per unit

… such that:

$$\text{Batch Value} = \text{Set Value} + \text{Run Value} \times \text{Batch Quantity}$$

… where 'Value' could be Cost, Time or Duration

Figure 6.1 illustrates how data in Table 6.5 can be normalised to any other batch quantity (within reason).

Plotting the data and fitting a best fit straight line through the data (see Volume III Chapter 4) allows us to determine that the 'Set Time' (time to set up the facility) is approximately 3 hours (i.e. this is the time it takes for zero order quantity), and thereafter each unit requires a 'Run Time' or processing time of around 2.2 hours.

Another quantity related value we might need to consider is the prior quantity produced as well as the quantity required to be produced both now and in the future. This allows us to take account of learning curve effects (both cost and schedule). Values can

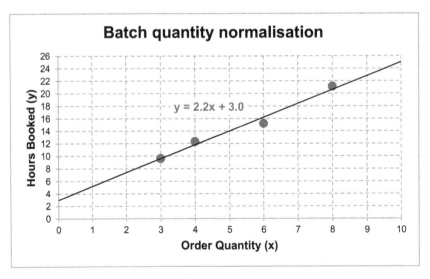

Figure 6.1 Example of Batch Quantity Normalisation

Table 6.5 Batch Quantity Normalisation

Works Order	Order Quantity	Hours Booked
700532	4	12.3
700889	3	9.6
701123	8	21.1
701345	6	15.2

be normalised to any point on the learning curve by factoring up or down the curve as appropriate. We will consider these in more depth in Volume IV Chapter 2.

We can also put throughput or output related adjustments into this category, i.e. the rate at which recurring activities are commenced or completed, e.g. Product Delivery Rate of 4 per month, Invoices Received per week, Quality Defects per 1,000 units.

6.3.3 Scale conversion – Fixed and Variable Factors

It is not unknown for estimators (or scientists, or engineers) to inadvertently introduce errors (*incredible but true, I'm afraid*) by mixing data that has been recorded or measured in different scale units, hence the importance of being clear in our spreadsheets in respect of the units being used (discussed in Chapter 3). The scale conversions we use may be fixed or variable factors.

> NASA lost the Mars Climate Orbiter (c.$125m) due to a failure to convert Imperial measurements to Metric measurements (Stephenson et al, 1999).

Fixed Factors, Rates or Ratios

… are used typically to normalise size data measured in different scale units, e.g. Imperial to Metric conversion, or in different units on the same scale system, e.g. feet and inches. Whilst in many cases it may be acceptable to use an approximation rule to convert between scales such as Imperial and metric, we don't have to. In some instances we can rely on Microsoft Excel instead to do all the jiggery-pokery using the function **CONVERT(number, from_unit, to_unit)**, saving us the trouble of remembering whether we multiply or divide by the conversion factor! This works for a variety of physical measures characterised by the following groups:

There are other functions within Excel that will convert specific units such as **DEGREES(radians)** and **RADIANS(degrees)**.

• weight/mass	• energy
• distance	• power
• time	• magnetism
• pressure	• temperature
• force	• liquid volume

Furthermore, it is very easy to lose the k or m when expressing a large value in thousands or millions. We know what we mean when we write or type it, but is it really clear to the next person who receives or reads it?

In these cases, the Factors, Rates and Ratios we use will always be fixed; they are axiomatic or inviolable relationships. Table 6.6 illustrates the normalisation procedure.

Variable Factors, Rates or Ratios

… are used typically to normalise data that is time dependent or ordered in some manner, and the Factor, Rate or Ratio used is dependent on the date/time or position in the sequence. Examples of adjustments bound to a point in time includes things such as escalation (inflation) or currency exchange rates. A learning curve might be used to adjust values which reduce as the cumulative quantity produced increases.

Sometimes, Variable Factors might be applied where there is a difference in performance or utilisation at a point or period in time. For example, what is the normalisation factor to convert man-hours to man-years? The answer is 'it depends on a number of variables and constants.' See the example in Table 6.7; the basic working week and work day are considered to be fixed but the absence, overtime and indirect or diversionary time are all variables over time.

An example of one of the most frequently used (and highly volatile), variable rate scale conversions is that of Foreign Exchange Rates (or ForEx), i.e. £ to γ, or $ to £.

Table 6.6 Scale Normalisation Using Fixed Factors

Part Number	Raw Data				Normalised Data			
	Length	UoM	Weight	UoM	Length	UoM	Weight	UoM
LNK-123987-X	9.7	in	1.55	lbm	24.638	cm	0.703068	kg
HNB-293668-Y-1	26.7	cm	783	g	26.7	cm	0.783	kg
HNA-005219-A	37.5	cm	1.124	kg	37.5	cm	1.124	kg
MMT-938464-N-3	15.25	in	39.3	ozm	38.735	cm	1.114136	kg

Unit of Measure Conversions based on:	Unit	UoM		Unit	UoM
	1	in	=	2.54	cm
	1	lbm	=	0.453592	kg
	1	g	=	0.001	kg
	1	ozm	=	0.02835	kg

Conversions perfomed with Microsoft Excel function **CONVERT**(*number, from_unit, to_unit*)
and the **Units of Measure (UoM)** are those recognised by that Excel Function

Table 6.7 Example of Variable Factors for Labour Utilisation

Planned labour utilisation for full time equivalent worker

						Hours	Net Hours	
Basic Working Year		52	weeks @	37.5	hours per week	1950	1950	Fixed
	Less Holidays	32	days @	7.5	hours per day	-240		Fixed
	Less Absence			2.67%	of Basic Hours	-52		Variable
	Plus Overtime			10.00%	of Basic Hours	195		Variable
Attendance Hours							1853	
	Less Diversions			9.60%	of Attendance Hours	-178		Variable
Direct Hours per Person per Year							1675	

Note: We should be careful not to get too sloppy when we are documenting our normalisation steps in our Basis of Estimate. Unless it is absolutely clear, rather than use the 'local' currency symbol such as £ or $, which neglects to advise us of the actual country we are talking about, we might want to consider using the less elegant but more definitive three letter ISO 4217 currency code, for example: GBP, USD, EUR etc. (ISO 4217:2008 Codes for the representation of currencies and funds).

Caveat augur

Even currency values (such as the Euro) that are generally considered to be common across a number of countries may need to be normalised across international boundaries for the same points in time. In the case of international collaborative projects that have contractual workshare arrangements agreed under fixed economic conditions expressed in terms of a common currency; they may require adjustment to take account of different rates of escalation in the countries involved thus distorting the current workshare!

That leads us on nicely to date-related changes.

6.3.4 Date or time related

Most people are aware of the concept of monetary inflation, where a pound or dollar today will buy you less than it did yesterday and more than it will do tomorrow (*figuratively speaking*), but there are instances where costs or prices fall over time (deflation). Some of the reducing patterns we might observe may be more volume related (see Section 6.3.2) than time related, but may appear to be time related as volumes often increase over time (*and overtime too perhaps*).

It is not just cost that varies over time, but other things also such as technological improvements. In 1965 Intel co-founder Gordon Moore observed that the number of integrated circuits on a micro-chip doubled every two years; he famously predicted that this trend would continue for the next ten years (Moore's Law). This empirical law seems to be just as relevant in principle today.

In order to deal with the practicalities of making adjustments in cost, performance etc., the estimator needs to be able to think in terms of the past and the future like '*wannabe time travellers*'. We will revisit this concept and explore the issues involved in more depth with examples in Section 6.4.

6.3.5 Life Cycle related

The estimates we may want to produce may be influenced by the stage of the product or project Life Cycle to which they relate. The Life Cycle may give us an indication of the level of cost, time or effort required (or the general level of efficiency), but also a source of potential risks and opportunities. We may find that these are often subjective judgements, but they can be guided by pseudo-objective measures, especially where the organisation has some formal Life Cycle review process or maturity gates in place. For instance, the UK Ministry of Defence utilises a Life Cycle by the acronym CADMID or CADMIT, reflecting the generic phases of:

- Concept
- Assessment
- Demonstration
- Manufacture or Migration
- In-Service
- Disposal or Termination

In comparing previous projects or products etc. that utilise different technologies, there are two dimensions we should consider. The first is that the technologies may be fundamentally different with different Estimating Relationships or metrics associated with them. However, we should also consider the impact of the state of development, or maturity, of a single technology on the entity that we are trying to estimate. As a useful means of assessing the relative position in the Technology Development Lifecycle, we might want to consider Technology Readiness Levels (TRLs) (Sadin et al, 1988).

Originally defined as seven levels, the current and perhaps most commonly used version has nine levels of technology development (Table 6.8) (NASA, 2004, p.DD-1). There are slight differences in the definitions between agencies and industries using the concept (and the detail), but this does not detract from the use of the concept.

Although TRLs were originally developed for Technology Development only, we might want to consider including a tenth stage for an ageing or retiring technology to include practical considerations about obsolescence and maintainability (Table 6.9).

Similar concepts have been developed (DoD, 2011) in relation to the Level of Manufacturing Readiness (Table 6.10).

As with the TRLs, we can extend the concept behind MRLs to take account of production end-of-line effects (Table 6.11) i.e. the impact due to shortages (stock outs), 'work arounds' and the dreaded problem of obsolescence (*hands up all those who would like to make obsolescence a thing of the past*).

Table 6.8 NASA Technology Readiness Levels (2004)

Level	Technology Readiness Description
TRL 1	Basic principles observed and reported
TRL 2	Technology concept and/or application formulated
TRL 3	Analytical and experimental critical function and/or characteristic proof-of-concept
TRL 4	Component and/or breadboard validation in laboratory environment
TRL 5	Component and/or breadboard validation in relevant environment
TRL 6	System/subsystem model or prototype demonstration in a relevant environment (ground or space)
TRL 7	System prototype demonstration in a space environment
TRL 8	Actual system completed and 'Flight qualified' through test and demonstration (ground or space)
TRL 9	Actual system 'Flight proven' through successful mission operations

Table 6.9 Potential Extension of Technology Readiness Levels to Reflect End-of-Life

TRL 10 or TRL X	Actual proven system towards the end of its intended life with increasing levels of obsolescent parts

Table 6.10 US DoD Manufacturing Readiness Levels (2011)

Level	Manufacturing Readiness Description
MRL 1	Basic manufacturing implications identified
MRL 2	Manufacturing concepts identified
MRL 3	Manufacturing proof of concept developed
MRL 4	Capability to produce the technology in a laboratory environment
MRL 5	Capability to produce prototype components in a production relevant environment
MRL 6	Capability to produce a prototype system or subsystem in a production relevant environment
MRL 7	Capability to produce systems, subsystems or components in a production representative environment
MRL 8	Pilot line capability demonstrated. Ready to begin low rate production
MRL 9	Low Rate Production demonstrated. Capability in place to begin Full Rate Production
MRL 10	Full Rate Production demonstrated and lean production practices in place

**Table 6.11 Potential Extension of Manufacturing Readiness Levels
to Reflect End-of-Life**

Level	Manufacturing Readiness Description
MRL 11	Actual proven production capability towards the end of its planned orders with increasing levels of obsolescent parts and stock shortages and reduced rate of production

Note: Different industries (such as defence, automotive, oil and gas) and different economic regions (such as the USA, European Union) have different versions of these Technology and Manufacturing Readiness Levels, and we should check what is recognised and used within our own organisations and supply chain.

If we have data to support the impact of evolving between these various states of readiness, we can apply the appropriate rules (e.g. Factors, Rates or Ratios); failing that we can use them to segregate or separate historical data into like or similar states of readiness (*sorting out the apples from the oranges and the pears – if we don't, we may find a few banana skins in our path!*).

The idea of segregating similar from dissimilar data can be further generalised by looking at other key groupings.

6.3.6 Key groupings – Role related

In order to avoid ourselves the embarrassment of making inappropriate comparisons that may seem to be reasonable at a superficial level, we should ask ourselves whether the products or services should be segregated in accordance with their intended or designed purpose or role, and/or operating environment. For example:

> Can we sensibly compare the cost to design, develop and operate military aircraft that perform different roles, e.g. attack, reconnaissance or transport? (*The answer is generally 'no', by the way, unless we start drilling down into the lower levels of the Work Breakdown Structure.*)
> Can we compare the cost to manufacture a manned vehicle with an unmanned one? (*For certain elements, we can, but holistically we can't.*)
> How can governmental defence departments compare the acquisition and operating costs of Land Vehicles (Armoured Personnel Carriers, Tanks etc.), Naval Ships (Aircraft Carriers, Destroyers, Frigates), Submarines, Aircraft (Fixed Wing, Rotary Wing), Missiles, Satellite Communications? (*The answer, of course, is 'with some difficulty!'*)

Even for a single product, there may be differences in the nature of how costs are incurred based on how the product is being used. For instance, taking any one of the four military services, Army, Navy, Air Force or Marines, does the cost of operating the service differ between these three scenarios?

- a routine training mission 'at home'
- a peace keeping mission on deployment abroad
- in 'theatre', i.e. a war zone

Whilst all of the above are questions from the defence industry, we can pose similar questions for other industries such as oil and gas:

- Does the cost of constructing or operating an oil and/or gas marine platform differ between the annual seasons?
- Is there any fundamental difference between the cost to construct a Chemical Tanker and a Petrochemical Tanker? (*Assuming, of course, that we are making a like-for-like comparison of tankers, i.e. between road vehicles, or merchant ships, not a mix and match of the two – it wasn't meant to be a trick question.*)

If we have any doubt, or just do not know the answer, then we might consider asking a Subject Matter Expert, or if there isn't one to hand (*There's never one around when you need one, is there?*), we can try plotting the data and observing the underlying pattern. For instance, in terms of the latter question, Figure 6.2 suggests that whilst there

> **Deadweight:** The difference in weight between a loaded and unloaded ship due to its cargo, fuel, fresh water, ballast, provisions, passengers and crew.

seems to be a high degree of comparability between the two types of Tankers, using the Tanker Deadweight as the Primary Cost Driver, there is a distinct, but measurable difference between them at this somewhat superficial high level. (*If we want to know why there is a difference, then we really need to find where that SME is hiding.*)

6.3.7 Scope related (subjective)

The previous section is clearly aimed at differentiating the scope at a very high level (apples vs. oranges), but even if we select only apples, there will be other more specific differences, e.g. is this a Granny Smith, Cox Pippin, Golden Delicious or Pink Lady? (*Other apples are available… as they say.*)

We may need to make adjustments for differences in scope of work which might be caused by:

- Differences in relation to actuals relative to the inclusion or exclusion of 'Recurring' and 'Non-recurring' activities. Some items may include non-recurring costs, and others exclude them
- Design and development philosophies may differ in terms of traditional design and development versus the impact on design of a Design-To-Cost (DTC) or Design-For-Manufacture (DFM) approach, which could make non-recurring more expensive in comparison with recurring costs (manufacture and/or operation)

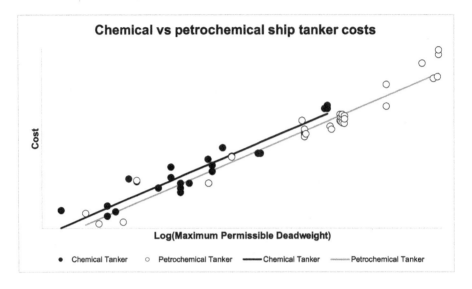

Figure 6.2 Chemical vs. Petrochemical Ship Tanker Costs

- Missing elements of work in one or more items relative to others
- Additional elements of work in one or more items relative to others
- More or less activity of a particular kind than is considered 'usual' for a product or service. For instance, this could relate to the level of integration activity or difficulty we might experience
- Differences in recorded values against the project, product or service due to changes in the Accounting system or conventions. An example of this for those of us contracting with the UK MoD would be a change in the activities that can be classified as being a direct chargeable activity as opposed to one which must be recovered through the overhead structure
- Changes in the condition of supply. For instance, the cost of a Spare may not always be the same as that of the Production equivalent because fastener holes may have been omitted to allow for alignment *in situ*. (*OK, fair point, holes are already an omission, what I mean is that we omit the activity that makes the hole!*)

To make these adjustments, we usually need to make a subjective judgement to add or remove elements, and these are most commonly done through factoring. However, if the granularity of the recording system is sufficient (e.g. in terms of the Work Breakdown Element selected), then it may be possible for us to make these normalisation adjustments in a more quantitative manner by estimating or measuring the difference.

Note (1): Recurring activities are generally associated with production activities and ongoing support to production or repetitive tasks. Non-recurring activities are those that are done once (or are intended to be done only once), usually in the project set-up or product development.

Note (2): Defined in each organisation's 'Questionnaire on the Method of Allocating Costs' (QMAC) and agreed with the MoD.

6.3.8 Complexity – Judgement related (subjective)

We may wish to segregate, rank or weight examples by the degree of complexity involved to complete a task etc. This inevitably requires some form of subjective judgement, even though we may use some form of quantitative measure as a guide. For instance, we may determine that the physical size of an item (either in terms of linear measurements or weight), is a good indicator of its complexity; we may decide also that small items can be compared with one another, and similarly large items might be compared together, but we cannot compare small with large items (or medium size ones, for that matter) without making some adjustment for their size.

Caveat augur

With so many different dimensions to consider, there is a risk of double counting their effects when normalising multiple entities. Take a step back and look at things holistically and ask yourself whether the total adjustment seems reasonable.

On the other hand, do not skimp or ignore the need to normalise data; it may seem tedious and time-consuming, but it is an essential step in the creative process we call estimating.

6.4 The estimator as a time traveller

There are some who believe in time travel; there are some that don't. As estimators, we just have to get our minds around the concept and imagine what it was or will be like in another time period. There are occasions when we have to make time stand still, turn it backwards, or forwards; the important thing is to avoid disorientation, and to know which way we are travelling.

In the main, making date-related adjustments is a special case of the Variable Factors, Rates and Ratios to which we alluded previously in Sections 6.3.3 and 6.3.4. It is often the case that for TRACEability (see Chapter 3) we will want to normalise data using reusable and credible indices.

It can be argued that the best sources of indices we could use are those that are generated by the organisation to which they relate. In reality, this is often only practical for our own organisations, unless there are contractual or legal obligations in place that allow access to those of our suppliers. In the majority of cases we will have to rely on industry 'averages' produced from reputable sources, often governmental bodies or agencies, but these could be international bodies, or professional institutions. The indices usually relate to cost and/or productivity movements. Many of these will publish monthly, quarterly and annual summaries of various indices such as:

- Consumer Price Index (CPI) or Retail Price Index (RPI)
  … which track the relative price change that consumers have to pay for a set of standard goods (*the proverbial shopping basket*), relative to a point in time, or Index Base Year
- Producer Price Indices (PPI) for a range of different industrial sectors and products
  … which track the relative price changes that those different industrial sectors charge for the goods they produce, relative to a point in time, or Index Base Year

If we extend the analogy with time travel a little further, to avoid disorientation when we travel forwards or backwards in time, it is recommended that we make adjustments relative to a consistent reference point in time.

6.4.1 Use of time-based indices 'Now and Then'

We have already determined that the estimator will need to normalise data to take account of historical economic changes such as price escalation, but for other business reasons the estimator may have to express the estimate in one or more ways going forward to include or exclude the effects of inflation or deflation. There are several terms used to describe these different scenarios (*some of them are not repeatable, so we will ignore them*). Unfortunately, there is often confusion between what the various terms mean that are used to describe values that have been adjusted to take account of inflation or deflation. Terms such as 'Current year' and 'Real Year' are often confused with one another, quite understandably. (*I blame the economists and the accountants.*) Consequently, we should always check that we are interpreting the terms in the same way as others in our customer-supplier-stakeholder family network.

Most of the confusion boils down to the flexibility of the English language and can be illustrated appropriately by the use of the words 'Now' and 'Then':

'Now' is synonymous with the present, such as 'Time Now', and we can use the word 'Then' to refer to a point in time in the past … or the future!

Let's consider, the 'Year' terms that cause the confusion, some of which are frequently used alternatives:

- Current Year   Does this mean 'Time Now' or that which was current at the time?
- Nominal Year   Implying that the year is not necessarily important?
- Then Year   Do we mean Then Past or Then Future?
- Outturn Year   Sounds straightforward
- Real Year   Does this mean Real as in Actual, or as in Real Terms?
- Constant Year   Sounds straightforward

The use of the word 'Current' may imply 'Time Now', whereas the generally accepted use in this context is in relation to the values that were current when they were observed or reported.

Definition 6.5   Current Year (or Nominal Year) Values

'Current Year Values' are historical values expressed in terms of those that were current at the historical time at which they were incurred. In some cases, these may be referred to as 'Nominal Year Values'.

We may wonder why these values are sometimes referred to as 'Nominal Year Values'. It may help to think of them as being transitory values that are nominally only representative at the time they are reported or observed.

A common mistake (*and one of which I have been guilty in the past*) is to think of these as being the values that were real at the time. Instead, the term 'Real Year Value' refers to something entirely different:

Definition 6.6   Constant Year (or Real Year) Values

'Constant Year Values' are values that have been adjusted to take account of historical or future inflationary effects or other changes, and are expressed in relation to the Current Year Values for any defined year. They are often referred to as 'Real Year Values'.

Perhaps the term 'Constant Year Value' is the more helpful one to use, as it implies better what is meant, and that relative to some defined 'Base Year' the impact of inflation has been removed to give a series of values in comparable 'real' terms. It may help us to think of 'Real Year Values' initially in the specific case of 'Time Now' as an expression of what is 'real' in today's 'Constant Year' economic environment, and that this will not change in the future if we keep the Base Year unchanged. Note: For Constant Year or Real Year Values to have any meaning, we must clearly define the 'Base Year' to which it is linked.

Our definition here refers to other predicted changes, not just inflation. For example, productivity may change over time due to technological changes, and we may need to normalise our data to take this into account.

To have any real meaning, the year in question for which Constant Year Values are being calculated, should be stated, and can reflect any year from the past, present or future. Talking about the future, how do we describe the future year values?

Definition 6.7 Then Year (or Outturn Year) Values

'Then Year Values' are values that have been adjusted to express an expectation of what might be incurred in the future due to escalation or other predicted changes. In some cases, these may be referred to as 'Outturn Year Values'.

It is unfortunate that the word '**Then**' can be used to imply both the past and the future, depending on in which direction we are looking. In the context of a 'Then Year', we must remember that here it is being used to look forwards not backwards. In many ways, 'Outturn Year' is a clearer definition, and is often preferred in some organisations.

To avoid the potential confusion, the following mnemonic may help:

'*What was **Current** is now **Past**, so let's look to the **Future Then**.*'

It is important to note that values may not always increase for 'Then Year Values'; deflationary values are not unknown (e.g. economic recessions, or technology maturity).

Note also that the Base Year in the context of a Constant or Real Year Value here should not be confused with the Base Year for a Table of Indices as discussed above, although the principle is the same insomuch as they are both fixed points of reference, and the latter is an example of the former. We can use an index for one Base Year to normalise values to another Constant Year Values for a different Base year.

Table 6.12 Escalation Using Different Economic Conditions

Description	Current Year			Base Year	Time Now	Then Years			
	2011	2012		2013	2014	2015	2016	2017	Total
Actual Escalation relative to previous year		2.2%		1.9%					
Assumed Escalation relative to previous year					2%	2%	2%	2%	
Escalation relative to 2011	1.000	1.022		1.041	1.062	1.083	1.105	1.127	
Pump Cost as Invoiced (Current Year - 2011)	£1,234								
Potential Pump Cost as Escalated (Rounded)	£1,234	£1,261		£1,285	£1,311	£1,337	£1,364	£1,391	
Deliveries per Year						10	20	20	50
Cost for 50 Pumps at Outturn ECs (Then Years - 2015 to 2017)						£13,370	£27,280	£27,820	£68,470
Cost for 50 Pumps at Constant Year (2014) EC values					↳	£13,110	£26,220	£26,220	£65,550
Cost for 50 Pumps at Contract Base Year (2013) EC values				↳		£12,850	£25,700	£25,700	£64,250

ECs = Economic Conditions

So that's probably as clear as mud then … or should that be 'now'! Perhaps an example (Table 6.12) will help us. Consider the following scenario:

- Suppose that 'Time Now' is early in 2014, and that we have an existing contract that requires a modified part to be supplied and fitted.
- A modified component for first delivery in 2015 requires the purchase of a new pump. The pump is a late-fit item.
- There is a requirement to deliver 10 components with replacement pumps in 2015, and a further 20 in each of the years 2016 and 2017.
- A similar pump for a comparable product cost £ 1,234 based on a June 2011 invoice we have found.
- We are required to estimate the pump replacement cost in the Contract's Base Year of 2013 ECs … but the Finance Director wants to know the cost in this year's 2014 Economic Conditions, … and the projected cost at outturn as well.

The procedure we can follow would be along the lines of:

- The known cost of a pump as invoiced in 2011 (***Current Year***), can be normalised using known annual escalation or inflation rates through to 2013, (the '***Base Year***' here relates to the date of the Contract being modified).
- Thereafter, the estimated cost of the pump can be escalated by an assumed rate of annual inflation; in this case we have applied a constant annual rate, but it could be a variable rate.
- We can express the future costs based on either a '***Constant Year***' value equivalent to the '***Base Year***' (2013), or any other '***Constant Year***' value; in this case, we have used the '***Time Now***' Year of 2014.
- Finally, we also illustrate how we can express the future values in the projected '***Then Year***' values for 2015 to 2017.

In the example, the row labelled 'Escalation relative to 2011' holds an index that measures the likely movement in the price we can expect due to the cumulative increase or decrease in the purchasing power of money over time to take account of average inflation or deflation for that type of commodity, relative to the Base Year of 2011. In many cases, as estimators we will be considering cost or price indices in this way, but in a more general sense, we may also find ourselves using indices in relation to any quantifiable entity such as productivity.

Definition 6.8   Index

An index is an empirical average factor used to increase or decrease a known reference value to take account of cumulative changes in the environment, or observed circumstances, over a period of time.

Typically, an index would be presented as one of a series of such values over a period of time. The main attraction of using indices is that they provide us with an opportunity for consistency and TRACEability in our assumptions both within a single estimate, and across similar or linked estimates.

To save us all the trouble of calculating indices for things such as the general movement in retail prices, or to stop us squabbling about the '*price of fish*', there are many authoritative and useful (*those two adjectives don't always go hand-in-hand*) sources of indices in the public domain:

- The 'Office for National Statistics' (ONS) in the United Kingdom
- The 'Bureau of Labor Statistics' (BLS) in the United States of America (one of several government agencies producing statistics)
- The 'Australian Bureau of Statistics' (ABS)
- 'Statistics Canada' or 'Statistique Canada' (*depending on where our political allegiances lie*)
- 'Eurostat' from the European Commission for the European Union
- The Organisation for Economic Co-operation and Development (OECD)

These can all be found by searching on the internet. (*We could have given links here but there is a risk that they become outdated, and the one thing more annoying that a broken hyperlink is one that you have tediously typed in to find it is broken.*) Check out also any local idiosyncrasies within these specific sources on how their indices have been calculated (see the next section), and how they are intended to be used. For instance, surprisingly to some, there are differences in the meaning of the terms and how the indices should be used between the different services of the US Military (Air Force, Navy, Marine and Army)!

6.4.2 Time-based Weighted Indices

A weighted index is one that is calculated based on the relative proportions of its constituent elements.

Not all indices are weighted averages (we may generate 'pure', or unweighted indices internally within our organisations), but we can argue that almost all indices available from public sources are likely to be weighted average indices. There are special cases of weighted indices that some sources

> For instance, a *Price of Fish Index* may be based on the retail or wholesale prices of different types of fish for sale in different fish markets around the country, weighted by the volumes of each sold in those locations. This way, no single location, or expensive type of fish, will dominate the overall *Price of Fish Index*.

like the International Cost Estimating & Analysis Association's Cost Estimating Body of Knowledge may refer to as 'Composite Indices' (ICEAA, 2009). (*No, not Indices about Composite Materials – well, they could be, but not in a general sense.*) Unfortunately, there is not necessarily a common understanding or use of the term across industrial sectors or geographic boundaries. This term tends to be used to refer to the weighted average values across a number of different, distinct but associated indices, e.g. labour and material costs or prices for given products or services. We will discuss these in Section 6.4.5.

There are a number of ways that the weighted indices can be calculated (*sorry, you didn't really think that there would only be one, did you?*). In economics, there are a number of different approaches, the most common of which are the **Laspeyres** and **Paasche Price Indices**, after the two eminent German economists who devised them. Both indices recognise that both the price and the mix of different commodity items may change, but they differ in how these changes are treated.

Goodridge (2007) of the Office for National Statistics commented that the choice of index was '*fairly arbitrary*' and the decision was generally taken '*based on practicalities*', such as the availability of reliable data. Whilst a Laspeyres Index relies on current prices, a Paasche Index relies on current quantities, which, as Goodridge observed, are often more difficult to obtain than price data or historical quantities. (*If we think of it logically, there is a natural lag in the system – the price of an item would generally be set before someone committed to buy.*) Consequently, many (but not all) of the indices published by the UK ONS (and probably other similar authoritative bodies) are based on the Laspeyres approach in preference to that of Paasche. However, within a single organisation, access to current quantities is probably less troublesome and therefore, for bespoke indices, we have a choice of the type of index we use.

For indices that are based on a large sample of different (but related) commodities, the difference between the two types of indices are not considered to be too problematic for estimators (*i.e. the accuracy versus precision debate*) but where there are major swings in the relative quantities across the years, there will be a measurable difference. We can explore

Index	Formula-phobes ...	Formula-philes ...
Laspeyres	... compares Prices of commodities at a point in time with the equivalent Prices for the Index Base Period, based on the original quantities consumed at the Index Base Year	$$I_t = 100 \times \frac{\sum p_{c,t}\, q_{c,0}}{\sum p_{c,0}\, q_{c,0}}$$ Summed over the range of commodities, c
Paasche	... compares Prices of commodities at a point in time with the equivalent Prices for the Index Base Period, based on the quantities consumed at the current point in time in question	$$I_t = 100 \times \frac{\sum p_{c,t}\, q_{c,t}}{\sum p_{c,0}\, q_{c,t}}$$ Summed over the range of commodities, c

Where:

$p_{c,0}$ = price of commodity, c, at time 0

$q_{c,0}$ = quantity of commodity, c, at time 0

$p_{c,t}$ = price of commodity, c, at time t

$q_{c,t}$ = quantity of commodity, c, at time t

Note: Whilst the examples here consider economic indices the principle of Laspeyres and Paasche can be applied to non-monetary indices.

Table 6.13 Example of Index Price and Quantity Data

Year	Price of Product x	Price of Product y	Quantity Sold of Product x	Quantity Sold of Product y	Quantity Sold of Both Products	Total Sales Revenue of Product x	Total Sales Revenue of Product y	Total Sales Revenue of Products x and y
2000	£10.99	£5.87	3073	6032	9105	£33,772	£35,408	£69,180
2001	£11.49	£6.37	3271	6072	9343	£37,584	£38,679	£76,262
2002	£11.49	£6.99	3794	5538	9332	£43,593	£38,711	£82,304
2003	£12.49	£7.99	4302	5273	9575	£53,732	£42,131	£95,863
2004	£12.99	£8.49	4722	5096	9818	£61,339	£43,265	£104,604
2005	£14.49	£8.99	5396	5067	10463	£78,188	£45,552	£123,740
2006	£14.99	£10.75	6037	4512	10549	£90,495	£48,504	£138,999

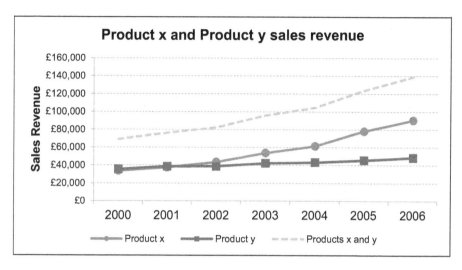

Figure 6.3 Example – Increasing Sales Revenue

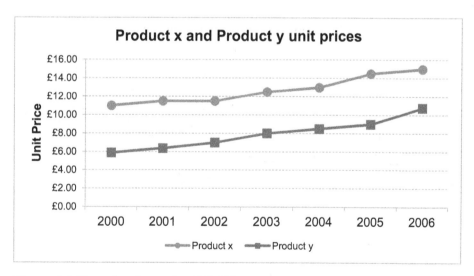

Figure 6.4 Example – Increasing Unit Prices

the difference in the two indices using the data in Table 6.13 for two excitingly named products **x** and **y**. Figures 6.3 to 6.5 also illustrate that whilst the total sales revenue of both products increases over time, as the unit prices increase, the demand for one product falls, whereas for the other it grows.

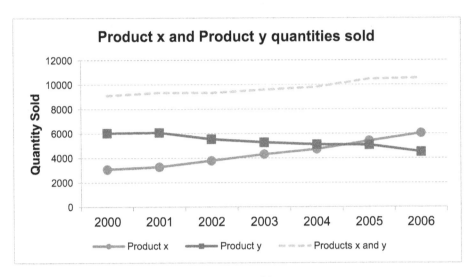

Figure 6.5 Example – Change in Quantities Sold

Table 6.14 Example of Laspeyres Weighted Price Index

Year	Price of Product x	Price of Product y	Quantity Sold of Product x	Quantity Sold of Product y	Sum of Base Year Quantities @ Current Year Prices	Sum of Base Year Quantities @ Base Year Prices	Laspeyres Weighted Index
2000	**£10.99**	**£5.87**	**3073**	**6032**	£69,180	£69,180	**100**
2001	£11.49	£6.37	3271	6072	£73,733	£69,180	106.6
2002	£11.49	£6.99	3794	5538	£77,472	£69,180	112.0
2003	£12.49	£7.99	4302	5273	£86,577	£69,180	125.1
2004	**£12.99**	**£8.49**	4722	5096	**£91,130**	**£69,180**	**131.7**
2005	£14.49	£8.99	5396	5067	£98,755	£69,180	142.8
2006	£14.99	£10.75	6037	4512	£110,908	£69,180	160.3

Table 6.15 Example of Paasche Weighted Price Index

Year	Price of Product x	Price of Product y	Quantity Sold of Product x	Quantity Sold of Product y	Sum of Current Year Quantities @ Current Year Prices	Sum of Current Year Quantities @ Base Year Prices	Paasche Weighted Index
2000	**£10.99**	**£5.87**	3073	6032	£69,180	£69,180	**100**
2001	£11.49	£6.37	3271	6072	£76,262	£71,591	106.5
2002	£11.49	£6.99	3794	5538	£82,304	£74,204	110.9
2003	£12.49	£7.99	4302	5273	£95,863	£78,231	122.5
2004	£12.99	£8.49	4722	5096	£104,604	£81,808	**127.9**
2005	£14.49	£8.99	5396	5067	£123,740	£89,045	139.0
2006	£14.99	£10.75	6037	4512	£138,999	£92,832	149.7

We can calculate the Laspeyres and Paasche Indices in Tables 6.14 and 6.15; the difference is illustrated in Figure 6.6.

The cells highlighted in bold in Tables 6.14 and 6.15 illustrate the data used in calculating the indices for 2004. Both the Laspeyres and Paasche Indices assume a Base Year Index of 100 at year 2000:

- The Laspeyres Index uses the Base Year Quantities Sold for 2000 invariably as the weighting values, applying them to the Current Year Prices for 2004 in comparison with the Base Year Prices for 2000. It multiplies the ratio of the two by the Base Year Index of 100. For example, the index for 2004 is derived as:

$$131.7 = 100 \times \frac{91130}{69180} = 100 \times \frac{(12.99 \times 3073 + 8.49 \times 6032)}{(10.99 \times 3073 + 5.87 \times 6032)}$$

- In contrast, the Paasche Index uses the Current Year Quantities Sold as the weighting values, applying them to the Current Year Prices for 2004 in comparison with those Current Year weightings for 2004 applied to the Base Year Prices for 2000. It multiplies the ratio of the two by the Base Year Index of 100. For example, the index for 2004 is derived as:

$$127.9 = 100 \times \frac{104604}{81808} = 100 \times \frac{(12.99 \times 4722 + 8.49 \times 5096)}{(10.99 \times 4722 + 5.87 \times 5096)}$$

Figure 6.6 demonstrates that there can be a marked difference in the way that the two indices calculate the economic pattern. The data has been chosen deliberately to highlight the potential of a difference; if the prices of each commodity and the

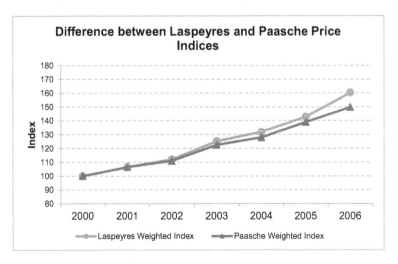

Figure 6.6 Example – Difference Between Laspeyres and Paasche Indices

associated quantities sold moved in the same direction, or there were many more commodities (rather than just these two), then the difference between the two Indices would be much less marked, hence why Goodridge can confidently say that the choice of system is largely arbitrary. The Laspeyres Index tends to overstate inflation (and *vice versa* for deflation) as it does not take account of changes in buying habits attributable to price changes, whereas the Paasche Index tends to understate inflation (and *vice versa* for deflation) for the reverse reason, i.e. by not acknowledging changes in price attributable to changes in buying habits.

If we are uncomfortable with this, and cannot be reassured by the soothing words of Peter Goodridge, there is a compromise we can explore. The **Fisher Price Index** (*seriously, that is what it is called – I'm not toying with words here*) takes the Geometric Mean of the Laspeyres and Paasche Indices. As this Fisher Index takes the best of worlds, and seeks to find the reality that lies between the optimistic and pessimistic views, it is sometimes referred to as the '**Ideal Index**'. *Note: 'Ideal' does not imply perfection!*

Table 6.16 builds on our previous examples in Tables 6.14 and 6.15, to derive the Fisher Weighted Index, by taking the Geometric Mean of the Laspeyres and Paasche Indices (i.e. the Square Root of their Product in this case).

Index	For the Formula-philes …	
Laspeyres	$I_t = 100 \times \dfrac{\sum p_{c,t} \, q_{c,0}}{\sum p_{c,0} \, q_{c,0}}$	
Paasche	$I_t = 100 \times \dfrac{\sum p_{c,t} \, q_{c,t}}{\sum p_{c,0} \, q_{c,t}}$	Summed over the range of commodities, c
Fisher	$I_t = 100 \times \sqrt{\dfrac{\sum p_{c,t} \, q_{c,0}}{\sum p_{c,0} \, q_{c,0}} \times \dfrac{\sum p_{c,t} \, q_{c,t}}{\sum p_{c,0} \, q_{c,t}}}$	
Where:	$p_{c,0}$ = price of commodity, c, at time 0	
	$q_{c,0}$ = quantity of commodity, c, at time 0	
	$p_{c,t}$ = price of commodity, c, at time t	
	$q_{c,t}$ = quantity of commodity, c, at time t	

Table 6.16 Example of Fisher Weighted Price Index

Year	Price of Product x	Price of Product y	Quantity Sold of Product x	Quantity Sold of Product y	Laspeyres Weighted Index	Paasche Weighted Index	Fisher Weighted Index
2000	£10.99	£5.87	3073	6032	100	100	100
2001	£11.49	£6.37	3271	6072	106.6	106.5	106.6
2002	£11.49	£6.99	3794	5538	112.0	110.9	111.4
2003	£12.49	£7.99	4302	5273	125.1	122.5	123.8
2004	£12.99	£8.49	4722	5096	131.7	127.9	129.8
2005	£14.49	£8.99	5396	5067	142.8	139.0	140.8
2006	£14.99	£10.75	6037	4512	160.3	149.7	154.9

For the Formula-phobes ... The index rule of thumb

(or perhaps we should call it the 'Index Finger Rule')

Rule of Thumb

When the prevailing trend in the indices is increasing, the following relationship tends to be true:

Laspeyres Index ≥ Fisher Index ≥ Paasche Index

... but when the prevailing trend is downward (e.g. deflation), then the reverse is true

6.4.3 Time-based Chain-linked Weighted Indices

As we have seen, neither the Laspeyres, Paasche nor Fisher Price Indices are perfect. Even though the 'Ideal' Fisher takes the middle road through the other two, it is fundamentally dependent on them. If current quantity data is difficult to obtain for the Paasche Index, then by default it is also difficult to obtain for the Fisher Index. To overcome the shortcomings of both the Laspeyres and the Paasche Indices in respect of overstating and understating inflation, the UK ONS (Goodridge, 2007), and many other authoritative sources of indices, use a technique called Chain-linking as a variation on the Laspeyres Index principally, but also in some instances, the Paasche Index.

Chain-linking is a technique which updates the weighting between commodities every year to reflect the movement in market conditions and preferences. In effect, it re-baselines the weightings every year, and uses them as the weightings for the following year. It then calculates the previous year's Index relative to the previous year rather than in relation to the original Base Year, which still has an Index of 100.

The difference between the two Indices is that in the Laspeyres Index the weightings are the previous year's quantities, whereas with the Paasche the weightings are the previous year's prices. We can compare the application of Chain-linking with our previous examples (Tables 6.17 and 6.18).

Index	For the Formula-philes ...

Chain-linked Laspeyres $\qquad I_t = I_{t-1} \times \dfrac{\sum p_{c,t}\ q_{c,t-1}}{\sum p_{c,t-1}\ q_{c,t-1}}$

$\left.\begin{array}{c} \\ \\ \\ \end{array}\right\}$ Summed over the range of commodities, c

Chain-linked Paasche $\qquad I_t = I_{t-1} \times \dfrac{\sum p_{c,t}\ q_{c,t}}{\sum p_{c,t-1}\ q_{c,t}}$

Where $\qquad I_0 = 100$

$\qquad p_{c,t}$ = price of commodity, c, at time t

$\qquad q_{c,t}$ = quantity of commodity, c, at time t

The cells highlighted in bold in Tables 6.17 and 6.18 illustrates the data used in calculating the indices for 2004. Both the Laspeyres and Paasche Chain-linked Indices commence with a Base Year Index of 100:

Table 6.17 Example of Laspeyres Chain-Linked Weighted Price Index

Year	Price of Product x	Price of Product y	Quantity Sold of Product x	Quantity Sold of Product y	Sum of Current Year Quantities @ Current Year Prices	Sum of Current Year Quantities @ Previous (Last) Year Prices	Paasche Chain-linked Weighted Index
2000	£10.99	£5.87	3073	6032	£69,180	£69,180	100
2001	£11.49	£6.37	3271	6072	£76,262	£71,591	106.5
2002	£11.49	£6.99	3794	5538	£82,304	£78,870	111.2
2003	£12.49	£7.99	4302	5273	£95,863	£86,288	123.5
2004	£12.99	£8.49	4722	5096	£104,604	£99,695	129.6
2005	£14.49	£8.99	5396	5067	£123,740	£113,113	141.8
2006	£14.99	£10.75	6037	4512	£138,999	£128,039	153.9

Table 6.18 Example of Paasche Chain-Linked Weighted Price Index

Year	Price of Product x	Price of Product y	Quantity Sold of Product x	Quantity Sold of Product y	Sum of Previous (Last) Year Quantities @ Current Year Prices	Sum of Previous (Last) Year Quantities @ Previous Year Prices	Laspeyres Chain-linked Weighted Index
2000	£10.99	£5.87	3073	6032	£69,180	£69,180	100
2001	£11.49	£6.37	3271	6072	£73,733	£69,180	106.6
2002	£11.49	£6.99	3794	5538	£80,027	£76,262	111.8
2003	£12.49	£7.99	4302	5273	£91,636	£82,304	124.5
2004	£12.99	£8.49	4722	5096	£100,651	£95,863	130.7
2005	£14.49	£8.99	5396	5067	£114,235	£104,604	142.8
2006	£14.99	£10.75	6037	4512	£135,356	£123,740	156.2

- The Laspeyres Chain-linked Index uses the Previous Year Quantities Sold as the weighting values, applying them to the Current Year Prices in comparison with the Previous Year Quantities and Prices. It multiplies the ratio of the two by the Previous Year's Index. For example, the index for 2004 is derived as:

$$130.7 = 124.5 \times \frac{100651}{95863} = 124.5 \times \frac{(12.99 \times 4302 + 8.49 \times 5273)}{(12.49 \times 4302 + 7.99 \times 5273)}$$

- In contrast, the Paasche Chain-Linked Index uses the Current Year Prices as the weighting values, applying them to the Current Year Quantities in comparison with the Previous Year Quantities at the Current Year Prices. It multiplies the ratio of the two by the Previous Year's Index. For example, the index for 2004 is derived as:

$$129.6 = 123.5 \times \frac{104604}{99695} = 123.5 \times \frac{(12.99 \times 4722 + 8.49 \times 5096)}{(12.49 \times 4722 + 7.99 \times 5096)}$$

In both cases of Laspeyres and Paasche (and therefore, by default Fisher also) the net result is that the indices move much close together as illustrated in Figure 6.7, the principal difference is that the weightings are only ever 'out' by a single year:

Figure 6.8 compares basic Laspeyres and Paasche Indices with their chain-linked versions.

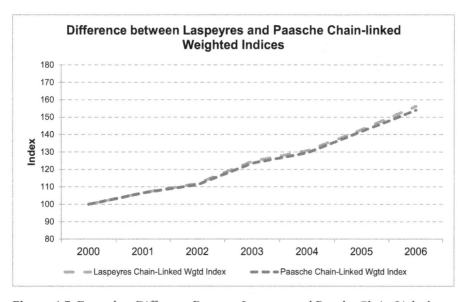

Figure 6.7 Example – Difference Between Laspeyres and Paasche Chain-Linked Weighted Indices

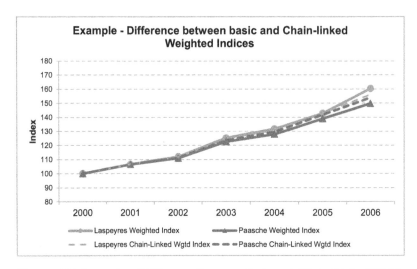

Figure 6.8 Example – Difference Between Basic and Chain-Linked Weighted Indices

For completeness, the Fisher Chain-linked Index is illustrated in Table 6.19. The main difference here is that it merely takes the geometric mean of the Laspeyres and Paasche Chain-linked Indices. It doesn't have to chain-link directly itself, but if we did we would get the same result – Chain-linking a Fisher Index is an unnecessary complication.

Table 6-19 Example of Fisher Chain-Linked Weighted Price Index

Year	Price of Product x	Price of Product y	Quantity Sold of Product x	Quantity Sold of Product y	Laspeyres Chain-linked Weighted Index	Paasche Chain-linked Weighted Index	Fisher Chain-linked Weighted Index
2000	£10.99	£5.87	3073	6032	100	100	100.0
2001	£11.49	£6.37	3271	6072	106.6	106.5	106.6
2002	£11.49	£6.99	3794	5538	111.8	111.2	111.5
2003	£12.49	£7.99	4302	5273	124.5	123.5	124.0
2004	£12.99	£8.49	4722	5096	130.7	129.6	130.2
2005	£14.49	£8.99	5396	5067	142.8	141.8	142.3
2006	£14.99	£10.75	6037	4512	156.2	153.9	155.0

For the Formula-philes: Chain-linking Fisher Indices

Consider a series of n Laspeyres Chain-linked Indices, L_1, L_2, ... L_n, and the corresponding Paasche Chain-linked Indices, P_1, P_2, ... P_n

Let l_n be the factor that adjusts the previous
Laspeyres Index, L_{n-1} to the latest Index L_n $\qquad$ $L_n = l_n L_{n-1}$ (1)

Let p_n be the factor that adjusts the previous
Paasche Index, P_{n-1} to the latest Index P_n $\qquad$ $P_n = p_n P_{n-1}$ (2)

The Fisher Index F_n is the Square Root of the
Product of the Laspeyres and Paasche
Indices, L_n and P_n $\qquad$ $F_n = \sqrt{L_n P_n}$ (3)

Substituting (1) and (2) in (3): $\qquad$ $F_n = \sqrt{l_n L_{n-1} p_n P_{n-1}}$ (4)

Simplifying (4): $\qquad$ $F_n = \sqrt{l_n p_n} \sqrt{L_{n-1} P_{n-1}}$ (5)

Simplifying (5) using (3): $\qquad$ $F_n = \sqrt{l_n p_n} F_{n-1}$

... which is the equivalent of Chain-linking the Fisher Index using the Square Root of the Laspeyres and Paasche Chain-link Factors. In other words, the Chain-link Factor for a Fisher Chain-linked Weighted Index is the Geometric Mean of the corresponding Laspeyres and Paasche Chain-link Factors.

6.4.4 The doubling rule for escalation

Rule of Thumb

Sometimes, in periods of steady state inflation or interest payments, (or at least an assumption of steady state average inflation or interest), it is useful to know how long it takes for a value to double in price (or halve in value). If we had access to a calculator, we could work it out quite quickly. Failing that, there is a useful Rule of Thumb, referred to as the 'Rule of 70' or 'Rule of 72'. In reality, it may be better to refer to it as the 'Rule of 70/72'.

For the Formula-phobes: How does the Rule of 70/72 work?

70 can be divided easily by the integers 1, 2, 5, 7, 10 and 14

(By 'easily' we mean 'in our heads' – anything is easy with a calculator)

(Continued)

72 can be divided easily by the integers 1, 2, 3, 4, 6, 8, 9 and 12

Between the two 'Rule Numbers', they can be divided by every integer from 1 to 14 with the exception of 11 and 13. As we are approximating, we can interpolate for the missing two between the Rule result for 10, 12 and 14 to give us $6^1/_2$ and $5^1/_2$.

For escalation or interest rates of 5% or less, the Rule of 70 is more accurate, whereas for 5% and above the Rule of 72 is the better.

For example:

- With a 5% escalation rate, we can divide 70 by 5 (stripping off the %) to get 14
- It will take approximately 14 years for a constant escalation rate to double the value

(1+5%) multiplied by itself 14 times (i.e. raised to the power of 14) equals 1.98, which is nearly double

Rule of Thumb

We can easily work out a similar Rule of Thumb, for how long it takes at a steady state to increase a value by 50%. This is the Rule of 40/42. Both the 70/72 and 40/42 Rules are illustrated in Table 6.20.

Table 6.20 The Rule of 70/72 and the Rule of 40/42

Number of Years to Double Value			Number of Years to Increase by 50%			
Annual % Escalation	Approximation		True Value (Rounded)	Annual % Escalation	Approximation		True Value (Rounded)
	Rule of 70	Rule of 72			Rule of 40	Rule of 42	
1	70	72	69.7	1	40	42	40.7
2	35	36	35.0	2	20	21	20.5
3		24	23.4	3		14	13.7
4	17.5	18	17.7	4	10		10.3
5	14	14.4	14.2	5	8	8.4	8.3
6		12	11.9	6		7	7.0
7	10		10.2	7		6	6.0
8		9	9.0	8	5		5.3
9		8	8.0	9			4.7
10	7	7.2	7.3	10	4	4.2	4.3
11			6.6	11			3.9
12		6	6.1	12			3.6
13			5.7	13			3.3
14	5		5.3	14		3	3.1

For the Formula-philes: Why do the Rules of 70/72 and 40/42 work?

Consider an Index, I_i, based on a steady state rate of growth of $x\%$ per annum, which after n years reaches the value N. The index base value is 100 at year 0.

The Index can be expressed as:
$$I_i = 100\left(1 + \frac{x}{100}\right)^i \quad (1)$$

After n years the Index $= N$
$$N = 100\left(1 + \frac{x}{100}\right)^n \quad (2)$$

Re-arranging and taking the Natural Log of (2):
$$\ln\left(\frac{N}{100}\right) = n\ln\left(1 + \frac{x}{100}\right) \quad (3)$$

Re-arranging (3):
$$n = \frac{1}{\ln\left(1 + \frac{x}{100}\right)}\ln\left(\frac{N}{100}\right) \quad (4)$$

The Natural Log of a value plus one can be expressed in the form of the Newton-Mercator Series (Havil, 2003):
$$\ln(1 + a) = a - \frac{a^2}{2} + \frac{a^3}{3} - \frac{a^4}{4} + \dots \quad (5)$$

For small values of $a < 1$
$$a\left(1 - \frac{a}{2}\right) < \ln(1 + a) < a \quad (6)$$

If $a < 15\%$ (inflation, interest is usually less than this):
$$0.925a < \ln(1 + a) < a \quad (7)$$

Inverting the inequality (7):
$$\frac{1}{a} < \frac{1}{\ln(1 + a)} < \frac{1000}{925a} \quad (8)$$

Substituting $a = \frac{x}{100}$ and the solution for n from (4):
$$\frac{100}{x}\ln\left(\frac{N}{100}\right) < n < \frac{100000}{925x}\ln\left(\frac{N}{100}\right) \quad (9)$$

Let N = 200 in (9) (i.e. double the Base Index):
$$\frac{100}{x}\ln(2) < n < \frac{100000}{925x}\ln(2) \quad (10)$$

Simplifying (10):
$$\frac{69.3}{x} < n < \frac{74.9}{x} \quad (11)$$

Alternatively, let N = 150 in (9) (i.e. 1.5 times the Base Index):
$$\frac{100}{x}\ln(1.5) < n < \frac{100000}{925x}\ln(1.5) \quad (12)$$

(Continued)

Simplifying (12):

$$\frac{40.5}{x} < n < \frac{43.8}{x} \qquad (13)$$

... leading to the Rules of 70/72 and 40/42

6.4.5 Composite Index: Is that not just a Weighted Index by another name?

The majority of indices that are available in the public domain are weighted indices based on a range of 'commodities' or 'things' within the scope of the particular index in question, e.g. price of leather products. It is often useful to develop indices that span a number of other indices to generate a 'Composite Index'. For instance, an organisation may want to know the overall increase in costs of its products including internal labour and bought out labour and non-labour (e.g. materials).

Whilst there is no commonly agreed or consistently used term across all industries, or organisations within an industry, there are some respected groups, such as the International Cost Estimating and Analysis Association (ICEAA), who refer to an Index compiled from a weighted average of other distinct Indices as a 'Composite Index'. We will follow their lead.

Definition 6.9   Composite Index

A Composite Index is one that has been created as the weighted average of a number of other distinct Indices for different commodities.

Table 6.21 provides an example using a fixed weighting. However, if we are to compile Composite Indices, we should consider whether the basis is to be Laspeyres, Paasche or both (i.e. Fisher), and whether we want to Chain-link them. The difficulty we will have is that often the quantities will be expressed in units that cannot be compared other than by a percentage weighting, and these are potentially influenced by the movement in the associated component Indices (*a bit of a circular argument*). The default is to use a fixed quantity-based weighting, which implies a standard Laspeyres Weighted Index.

We can easily create these Composite Indices within Microsoft Excel with the **SUMPRODUCT(array1, array2, ...)** function, in which *array1* would be the range of weighting factors (totalling 100%) and *array2* is the range of indices across the constituent elements for a single time period (*or vice versa*).

Table 6.21 Example of a Composite Index

	Proportion of Material Content				
	Steel	Copper	Aluminium	Electronic	Total
Weighting	20%	35%	30%	15%	100%
	Published Producer Price Index				Composite Index
	Steel	Copper	Aluminium	Electronic	
2007	100.0	100.0	100.0	100.0	100.0
2008	102.1	101.9	102.3	101.1	101.9
2009	104.9	103.6	105.0	101.4	104.0
2010	108.6	107.5	108.9	101.9	107.3
2011	112.2	110.3	112.8	102.5	110.3
2012	113.4	112.8	113.8	102.4	111.7
2013	114.8	113.5	115.4	101.9	112.6

6.4.6 Using the appropriate appropriation approach

This section could easily have been called 'Watch out there are tripwires about!', but the alliteration sounded better.

Some institutions, quite naturally want to know what their budget requirements are likely to be in today's 'Current Year' money values, but also what they need to be budgeting for in future years in 'Then Year' values.

Consider the situation where we are being asked to request funding today for values over a number of years going forward, but we are being asked to quote the budget request inclusive of escalation. Apart from the difficulty of predicting future rates of inflation, we may not know the precise start date of a project, but have a reasonable view of the spend profile between the start and finish based on prior similar projects.

Let's consider the theoretical spend profile in Figure 6.9 over a four-year period assuming 'Constant Year' values (i.e. no inflation).

If we make a working assumption that this will be subject to an average of 3% annual escalation in terms of the budget that we will actually require across the four years, then we can generate the results in Table 6.22.

We can calculate the Equivalent 'Outturn' Budget required at 'Then Year' Economic Conditions (ECs) by factoring up the 'Base Year' Values for escalation. This is equivalent to multiplying the Total Budget at 'Base Year' ECs by a **Weighted Arithmetic Mean** of the Escalation Factors, where:

- The Escalation Factors are the 'Then Year' Escalation Indices divided by 'Base Year' Escalation Index
- The weightings are the annual percentages of the 'Base Year' Budget

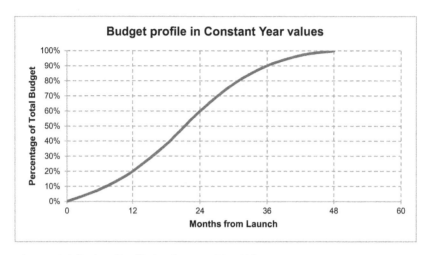

Figure 6.9 Budget Profile in Constant Year Values

Table 6.22 Factor Up from 'Base Year' to 'Then Year'

Total Budget (£k) >>	£800.00 k	at Base Year Economic Conditions			
Average Annual Escalation >>		3%			
Year	% of Total Budget in Base Year ECs	Budget Required in Base Year ECs	Escalation Index relative to Base Year	Annual Budget in Then Year ECs	% of Total Budget in Then Year ECs
1	20%	£160.00 k	100	£160.00 k	19.2%
2	40%	£320.00 k	103.0	£329.60 k	39.6%
3	30%	£240.00 k	106.1	£254.62 k	30.6%
4	10%	£80.00 k	109.3	£87.42 k	10.5%
Total	100%	£800.00 k	↗	£831.63 k	100%

Conversely in Table 6.23, we can 'reverse engineer' an Equivalent 'Outturn' Budget back to 'Base Year' Values by factoring down from the 'Then Year' Values. This is equivalent to dividing the Total Budget at 'Then Year' ECs by a **Weighted Harmonic Mean** of the Escalation Factors (see Volume II Chapter 2), where:

- The Escalation Factors are the 'Then Year' Escalation Indices divided by 'Base Year' Escalation Index
- The weightings are the annual percentages of the 'Then Year' Budget

When we are dealing with 'Base Year' or any 'Constant Year' values, life is more straight-forward. Any shift to the right or left in the programme overall over a fixed duration

Table 6.23 Factor Down from 'Then Year' to 'Base Year'

Total Budget (£k) >>	£831.63 k	at Then Year Economic Conditions		
Average Annual Escalation >>		3%		
Year	% of Total Budget in Then Year ECs	Annual Budget in Then Year ECs	Escalation Index relative to Base Year	Budget Required in Base Year ECs	% of Total Budget in Base Year ECs
1	19.2%	£160.00 k	100	£160.00 k	20.0%
2	39.6%	£329.60 k	103.0	£320.00 k	40.0%
3	30.6%	£254.62 k	106.1	£240.00 k	30.0%
4	10.5%	£87.42 k	109.3	£80.00 k	10.0%
Total	100%	£831.63 k	↘	£800.00 k	100%

For the Formula-philes: Appropriation profile rules ... OK?

Consider a time-phased cost profile over n time periods. Let b_1, b_2, ... b_n represent the series of values expressed in 'Base Year' Economic Conditions, and let t_1, t_2, ... t_n represent corresponding values in 'Then Year' Economic Conditions based on the range of Escalation Indices, E_1, E_2, ... E_n. (Typically $E_1 = 100$)

Let ε_i be the Escalation Factor at time period i relative to the Base Year Index, E_1 such that:

$$\varepsilon_i = \frac{E_i}{E_1} \qquad (1)$$

By definition, the 'Then Year' cost element, t_i, in time period i can be expressed in relation to the equivalent 'Base Year' cost element, b_i, multiplied by the Escalation Factor:

$$t_i = b_i \varepsilon_i \qquad (2)$$

The Total Cost in 'Base Year' values, B_{Tot}, is the sum of its profile across the time periods:

$$B_{Tot} = \sum_{i=i}^{n} b_i \qquad (3)$$

The Total Cost in 'Then Year' values, T_{Tot}, is the sum of its profile across the time periods:

$$T_{Tot} = \sum_{i=i}^{n} t_i \qquad (4)$$

Substituting (2) in (4):

$$T_{Tot} = \sum_{i=i}^{n} b_i \varepsilon_i \qquad (5)$$

By definition, the Weighted Arithmetic Mean, WAM_b, of the Escalation Factors, $\varepsilon_1 ... \varepsilon_n$ using the profile weightings, $b_1 ... b_n$ is:

$$WAM_b = \frac{\sum_{i=i}^{n} b_i \varepsilon_i}{\sum_{i=i}^{n} b_i} \qquad (6)$$

(Continued)

Substituting (5) and (3) in (6):

$$WAM_b = \frac{T_{Tot}}{B_{Tot}} \qquad (7)$$

Re-arranging (7), we can express the Total 'Then Year' Cost as the Total 'Base Year' Cost multiplied by the Weighted Arithmetic Mean of the Escalation Factors over the 'Base Year' profile value:

$$T_{Tot} = WAM_b \, B_{Tot}$$

By definition, the Weighted Harmonic Mean, WHM_t, of the Escalation Factors, $\varepsilon_1 \ldots \varepsilon_n$ using the profile weightings, $t_1 \ldots t_n$ is:

$$WHM_t = \frac{\sum_{i=i}^{n} t_i}{\sum_{i=i}^{n} \frac{t_i}{\varepsilon_i}} \qquad (8)$$

Substituting (4) and (2) in (8):

$$WHM_t = \frac{T_{Tot}}{\sum_{i=i}^{n} b_i} \qquad (9)$$

Substituting (3) in (10):

$$WHM_t = \frac{T_{Tot}}{B_{Tot}} \qquad (10)$$

Re-arranging (11), we can express the Total 'Then Year' Cost as the Total 'Base Year' Cost divided by the Weighted Harmonic mean of the Escalation Factors over the 'Then Year' profile values:

$$B_{Tot} = \frac{T_{Tot}}{WHM_t}$$

will not affect our budget profile, provided the 'Base Year' Economic Conditions do not alter. However, if we are dealing with 'Then Year' values, any shift to the right or left in the programme overall will fundamentally change the Equivalent 'Outturn' Budget due to the change in the weightings and Escalation Factors, as illustrated in Table 6.24 and Figure 6.10.

In this case, we have assumed a six-month delay in the programme overall, resulting in an increase in the calculated 'Then Year' values of approximately 1.5%. Of course, any change in the programme duration will render both our 'Base Year' and 'Then Year' profiles as redundant as they are only valid in the context of the defined programme.

So, why are these 'Appropriation Profiles' so important?

Typically, we may use a budget or price proposal based on a fixed reference year or 'Base Year' where we have a contractual agreement that allows us to vary the price according to an agreed escalation formula based on what eventually comes to pass in the future (*i.e. the client or customer carries the inflation risk*).

Alternatively, if there is no option within the contract to vary the price according to an escalation formula, we should develop a budget or price proposal based on our view of the 'Then Year' or 'Outturn' Values (*i.e. the contractor carries some of the inflation risk*).

Table 6.24 Cumulative Effect of Programme Delay

Total Budget (£k) >>	£800.00 k	at Base Year Economic Conditions		
Average Annual Escalation >>		3%		
Year	% of Total Budget in Base Year ECs	Budget Required in Base Year ECs	Escalation Index relative to Base Year	Annual Budget in Then Year ECs	% of Total Budget in Then Year ECs
1	8%	£64 k	100	£64.00 k	7.6%
2	30%	£240 k	103.0	£247.20 k	29.3%
3	40%	£320 k	106.1	£339.49 k	40.2%
4	19%	£152 k	109.3	£166.09 k	19.7%
5	3%	£24 k	112.6	£27.01 k	3.2%
Total	100%	£800 k	↗	£843.79 k	100%

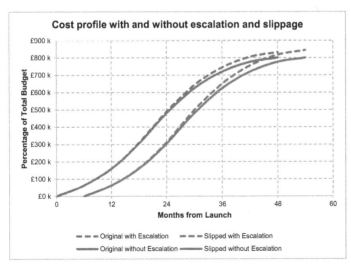

Figure 6.10 Cost Profile with and without Escalation and Slippage

6.4.7 Use of time as an indicator of other changes

In some circumstances, we can use time as an indicator of other changes, not just economic or financial inflation. We can use it where we believe that there is a natural progression through time in some cost driver. These can be positive or negative trends. For instance, we may consider that the impact of new working practices across an industry occurs incrementally over time.

If we are using Microsoft Excel, we can use this time-based indicator as either a date, or as a relative passage of time since a point in time, e.g. Month 1, Month 2, Month 3.

> If we are using Excel, these are equivalent, as Excel calculates and stores a date as the number of days since 1st January 1900.
>
> *Hmm, so that was the beginning of time! No, rewind, it can't be, most of the Maths and Stats we are using in this book pre-date that, so they wouldn't exist. OK, it must be just some arbitrary convenient reference point rather than some startling scientific revelation uncovered by Microsoft.*

We might consider time to be a useful indicator in some of the following:

- Working Practices/Efficiency Improvements
- Technology
- Wear and Tear
- Legislative Controls

Note: this is not meant to be a definitive list, just a prompt to consider other 'underlying', possibly somewhat less tangible, drivers.

In case some of us are thinking '*Can we save some time by combining cost escalation and technology advancements into a single time-based escalation measure?*' then in theory, the answer is '*Yes,*' but it is better, wherever possible, to normalise our data first for the inflationary effects before considering the underlying impact of the more intangible Drivers. It would be both risky and naive to assume that their combined effects would be any smoother in reality, than we would get by applying them individually.

6.5 Discounted Cash Flow – Normalising investment opportunities

6.5.1 Discounted Cash Flow – A form of time travel for accountants

No, nothing to do with knock-down bargain prices, although future revenues are reduced!

Discounted Cash Flow Analysis is another example of a forward-looking normalisation technique that estimators and accountants are often called on to use. Discounting is

---

Definition 6.10 Discounted Cash Flow (DCF)

Discounted Cash Flow (DCF) is a technique for converting estimated or actual expenditures and revenues to economically comparable values at a common point in time by discounting future cash flows bay an agreed percentage discount rate per time period, based on the cost to the organisation of borrowing money, or the average return on comparable investments.

---

a technique we can use which adjusts Cash Flows forecasts to take account of the time value of investment opportunities. As the definition implies, DCF can be performed on actual data from a retrospective assessment or audit of the return on an investment, but it has probably been performed previously as part of the initial investment appraisal and approval, which is forward-looking.

The principle of how the technique of Discounting works is the same as that for escalation; however, the Discount Rate applied is *not* the same as an inflation rate. Instead, we need to consider Discounting as a technique used primarily for comparing alternative Investment Opportunities, or cost-benefit trade-offs.

> We can justifiably claim that there is a link, albeit a little tenuous, between Discounting and Inflation insomuch that Investment Opportunities are linked with inflationary pressures but the two are not perfectly correlated by any means.

Definition 6.11 Discount Rate

The Discount Rate is the percentage reduction used to calculate the present-day values of future cash flows. The discount rate often either reflects the comparable market return on investment of opportunities with similar levels of risk, or reflects an organisation's Weighted Average Cost of Capital (WACC), which is based on the weighted average of interest rates paid on debt (loans) and shareholders' return on equity investment.

The Discount Rate is the rate at which the value of money decays in relation to other investment opportunities (Figure 6.11):

By now, you'll not be surprised to hear that there is more than one approach to Discounting, and the choice is often down to Corporate Policy rather than any altruistic

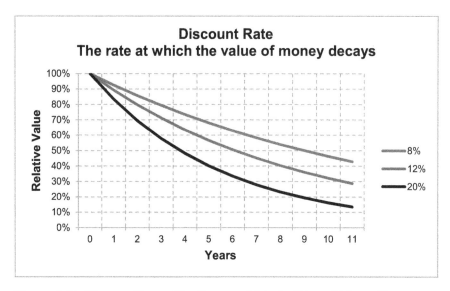

Figure 6.11 Discount Rate – The Rate at which the Value of Money Decays

reason. The choices tend to be, but are probably not restricted to, those summarised in Table 6.25.

From a purely practical point of view, the latter two situations in Table 6.25 in respect of the timing of transactions are only really pertinent in the case of a lengthy time period such as a year. In the majority of cases, a time period of a month is probably too short to '*split hairs*' between middle or end of month transactions and alternative investment opportunities.

The three main techniques for investment appraisal are Net Present Value (NPV), Internal Rates of Return (IRR) and Payback Period. (*There are others that we will not cover such as Adjusted Present Value or APV.*) They are all linked. It is possible, should we wish, to consider and apply these techniques over any granularity of time (e.g. month) but the most frequently used time period across industry is 'year'. Consequently, for the rest of this chapter, we will assume that the time period will be years.

The principle of making a comparison, using a consistent normalised approach is the key factor. Organisations will choose the model which best suits them.

Ideally, when comparing alternative investment opportunities, we should always look at the big picture in terms of utility and disutility. In other words, wherever practical, we should look at the cost of doing one thing and the cost of not doing something else, but eliminating those costs which are common to both alternatives. For example, in terms of a Make versus Buy decision, we should try to compare:

Cost of Making + Cost of Not Buying vs. Cost of Buying + Cost of Not Making

Table 6.25 Discounted Cash Flow – Some Alternative Approaches

Investment approved ...	Year 0	Year 0
Discounting commences ...	...for Values at Period 0	...for Values at Period 1
Implies that ...	Period 0 is the point of investment review and approval, and also of investment commencement	Period 0 is the point of investment review and approval and Period 1 is the point of investment commencement
Cash Flow transactions occur ... **... at the end of each period**	Discounting formula must be chosen to reflect the appropriate Discount Start Point (0 or 1)	
... in the middle of each period	Requires an adjustment to the period 'counter' to account for the half period offset from the Discount Start Point	

For instance, consider:

- Option to sub-contract work at a lower charging rate   => utility
- Sub-contracting lowers the potential internal throughput and may adversely affect the overall overhead cost recovery for the business   => disutility
- Option of not sub-contracting work may result in failing schedule internally due to capacity constraints   => disutility
- Option of sub-contracting may provide strategic leverage with the sub-contractor   => utility

6.5.2 Net Present Value (NPV)

Definition 6.12   Net Present Value (NPV)

The Net Present Value (NPV) of an investment is the sum of all positive and negative cash flows through time, each of which have been discounted based on the time value of money relative to a Base Year (usually the present year).

It is usually considered to be custom and practice to develop business cases for investment opportunities in Constant Year Values. Net Present Value is a technique we can use to take account of the value of future costs or benefits in relation to a Base Year, often the current year, or Time Now. (However, the important thing to stress is that any comparison of the return on investment between alternative investment opportunities is performed in relation to a common Base Year.)

For the Formula-phobes ... Net Present Value

Suppose we have an opportunity to invest in an internet business (dot.com) over three years that is expected to make around £ 1,250 return on our £ 10,000 capital based on current projections (thus implying a level of uncertainty, or even risk or opportunity).

If our alternative is to deposit £ 10,000 in a high interest savings account for a fixed term of three years with a guaranteed return of 4% per annum, then with Compound Interest, it would yield £ 1,248.64 in interest.

For the sake of £ 1.36, we would probably go for the 'safe' option of the Savings Account. However, if the Savings Account yielded 5% guaranteed interest per annum, then undoubtedly, we would go for the safe option, but if instead it only accrued 3% Interest per annum (or £ 927.27 in three years), we may be tempted to take the more riskier option of the internet investment, *and keep our fingers crossed*.

Net Present Value allows us to compare investment options over time

---

Each organisation is likely to have an expectation of a rate of return (*hurdle rate*) on any investment opportunity which may (*or may not*) be linked to the perceived level of risk involved in undertaking that investment, e.g. 8%, 12% or 20% return for low, medium or high risk projects. (*Note that these are for illustration purposes only and do not constitute a recommendation!*) This hurdle rate, when expressed as an average annual rate is the '*Discount Rate*', and is usually based on the Weighted Average Cost of Capital (WACC), which is a measure of how much an organisation must pay to borrow money from all sources (i.e. debt and equity), and how much risk the marketplace perceives there to be when investing in that organisation. Investors will expect a higher rate of return for higher risk investments.

NPV analysis discounts the value of the future costs and benefits accumulated each year by the Discount Rate. Typically we might compute the NPV in order to make a comparison:

i.   Between alternative investment opportunities (i.e. Project X vs. Project Y)

ii.   With a target minimum rate of return (the Discount Rate), i.e. does it create a positive
      NPV over the life of the project?

Consequently, any NPV assessment should only consider differences between the alternative investments, or changes to the current state, rather than the full cost to the business.
For example:

- A new manufacturing or construction process may require three people to operate
  it, but it is replacing one that requires four people. The full operating cost of the new
  process would require the cost of three people for the duration, but the saving that
  the investment brings is the cost of one person. NPV would use the latter
- Any actual costs incurred in evaluating an investment opportunity (e.g. proof of
  concept testing) should be ignored. They are sunk costs, and will exist irrespective
  of which future investment opportunity is pursued

For the Formula-philes … Net Present Value calculations: The options

Consider an investment opportunity over the following n years, relative to the
Base Year 0, which yields a series of net cash flows per annum: $C_0 \ldots C_n$.

The Net Present Value (NPV) can be calculated in one of three ways.

- Investment commences at the end of Year 0.
  Discounting commences at the end of Year 1
  Cash Flow transactions assumed to occur
  at the end of each Year

$$NPV = \sum_{t=0}^{n} \frac{C_t}{(1+\delta)^t} \quad (A)$$

- Investment commences at the start of Year 0.
  Discounting commences at the end of Year 0
  Cash Flow transactions assumed to occur at
  the end of each Year

$$NPV = \sum_{t=0}^{n} \frac{C_t}{(1+\delta)^{t+1}} \quad (B)$$

- Investment commences at the end of Year 0.
  Discounting commences at midpoint of Year 0
  Cash Flow transactions assumed to
  occur at Mid-Year Point

$$NPV = \sum_{t=0}^{n} \frac{C_t}{(1+\delta)^{t+0.5}} \quad (C)$$

Table 6.26 draws a comparison of the different approaches at our disposal. From this simple example, we can demonstrate that Option (A) is the least aggressive and Option (B) is
the most aggressive when it comes to discounting; Option (C), by using the assumption

Table 6.26 Discounted Cash Flow – Alternative Approaches

Discount Rate >	10%			
Year	Net Cash Flow	NPV Method (A)	NPV Method (B)	NPV Method (C)
0	-$1,000	-$1,000	-$909	-$953
1	$200	$182	$165	$173
2	$800	$661	$601	$630
3	$1,600	$1,202	$1,093	$1,146
Total	$1,600	$1,045	$950	$996

that all transactions occur on average in the middle of each year, is taking the '*halfway house*' between Options (A) and (B). Putting this into context, none of the assumed models are wholly representative of reality, as cash flows will occur at various points in each year, depending on the projects in question. We can create any of these three NPV formulae 'long hand' in Microsoft Excel, but only one of them can be reproduced directly using the inbuilt Excel function **NPV(*rate, list of values*)**; this is Formula (B) above. It is possible, of course, to be a little creative in how we present and use our data in order to utilise the NPV function in Excel for Option (A). To do this we have to ignore any transactions in Year 0 for the range used in the NPV function initially, and then to add them on afterwards to get the true NPV calculation we want.

Table 6.27 illustrates the process, assuming DCF Method Option A. Figure 6.12 presents this example pictorially. Later in this section we will discuss Payback Periods.

Often, we may use NPV in isolation, leaving us to compare the outturn merely against the benchmark of '*does it make financial sense*'?

- If the NPV is positive, the investment project is financially viable (i.e. it makes a positive return). A positive NPV represents the Profit or Return which the investment is predicted to make, expressed in Present Day Values relative to the return expected for the type of investment in question
- If the NPV is negative, the investment is not financially viable (i.e. it makes a negative return or loss) compared with expectations
- An NPV of Zero indicates that the investment Breaks Even

In the example, we can conclude that over the life of the project, the investment is financially viable as it returns a positive NPV. However, if the project fails to generate the forecast sales revenue in the Years 2020–2022, then the project would not have generated a positive NPV by 2019 to justify the investment.

One of the greatest benefits of the NPV technique is that it allows us to make a comparison between two investment opportunities (projects) over different timescales.

Table 6.27 Capital Investment Case – Discounted Cash Flow Analysis

NPV Method (A)	Investment & Operating Cost Delta					Discount Rate 14%	
Year	Capital Purchase Costs	Commissioning & Operational Costs	Projected New Sales Revenue	Projected Net Annual Cost Delta	Cumulative Cash Flow (at Base Year Values)	Present Value (Discounted)	Cumulative Present Value (Discounted)
0 2011	€ 0	€ 0	€ 0	€ 0	€ 0	€ 0	€ 0
1 2012	-€ 257,900	-€ 107,000	€ 0	-€ 364,900	-€ 364,900	-€ 320,088	-€ 320,088
2 2013	-€ 245,000	-€ 103,350	€ 50,000	-€ 298,350	-€ 663,250	-€ 229,571	-€ 549,658
3 2014		-€ 4,502	€ 150,000	€ 145,498	-€ 517,752	€ 98,207	-€ 451,451
4 2015	-€ 232,750	-€ 100,660	€ 200,000	-€ 133,410	-€ 651,162	-€ 78,989	-€ 530,441
5 2016		-€ 5,750	€ 250,000	€ 244,250	-€ 406,912	€ 126,856	-€ 403,585
6 2017		-€ 5,750	€ 300,000	€ 294,250	-€ 112,662	€ 134,056	-€ 269,529
7 2018		-€ 5,750	€ 300,000	€ 294,250	€ 181,588	€ 117,593	-€ 151,935
8 2019		-€ 5,750	€ 300,000	€ 294,250	€ 475,838	€ 103,152	-€ 48,783
9 2020		-€ 5,750	€ 300,000	€ 294,250	€ 770,088	€ 90,484	€ 41,701
10 2021		-€ 5,750	€ 300,000	€ 294,250	€ 1,064,338	€ 79,372	€ 121,073
11 2022		-€ 5,750	€ 300,000	€ 294,250	€ 1,358,588	€ 69,625	€ 190,698
Total	-€ 735,650	-€ 355,762	€ 2,450,000	€ 1,358,588	NPV >	€ 190,698	

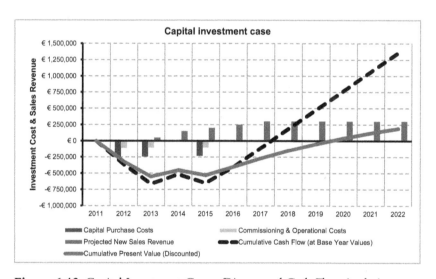

Figure 6.12 Capital Investment Case – Discounted Cash Flow Analysis

For instance, in Table 6.28 and Figure 6.13 we consider an alternative investment opportunity to the one we discussed in Table 6.27 and Figure 6.12. Here we are considering a lower level of investment, that starts later in time and generates less Sales Revenue. In order to make a fair comparison with the original investment case, we must begin our comparison in the same year, otherwise we will have applied different discounts to the same calendar years where the investments overlap.

In this particular case, the NPV of the project is positive, and is slightly higher than the earlier opportunity, and therefore it is the better opportunity.

Table 6.28 Alternative Capital Investment Case – Discounted Cash Flow Analysis

	NPV Method (A)	Investment & Operating Cost Delta					Discount Rate 14%	
	Year	Capital Purchase Costs	Commissioning & Operational Costs	Projected New Sales Revenue	Projected Net Annual Cost Delta	Cumulative Cash Flow (at Base Year Values)	Present Value (Discounted)	Cumulative Present Value (Discounted)
0	2011	€ 0	€ 0	€ 0	€ 0	€ 0	€ 0	€ 0
1	2012	€ 0	€ 0	€ 0	€ 0	€ 0	€ 0	€ 0
2	2013	-€ 203,450	-€ 48,750	€ 0	-€ 252,200	-€ 252,200	-€ 194,060	-€ 194,060
3	2014	-€ 102,300	-€ 24,900	€ 66,000	-€ 61,200	-€ 313,400	-€ 41,308	-€ 235,368
4	2015		-€ 9,576	€ 155,000	€ 145,424	-€ 167,976	€ 86,103	-€ 149,265
5	2016		-€ 9,576	€ 179,000	€ 169,424	€ 1,448	€ 87,994	-€ 61,272
6	2017		-€ 9,576	€ 179,000	€ 169,424	€ 170,872	€ 77,187	€ 15,916
7	2018		-€ 9,576	€ 179,000	€ 169,424	€ 340,296	€ 67,708	€ 83,624
8	2019		-€ 9,576	€ 179,000	€ 169,424	€ 509,720	€ 59,393	€ 143,017
9	2020		-€ 9,202	€ 172,000	€ 162,798	€ 672,518	€ 50,062	€ 193,078
10	2021							
11	2022							
	Total	-€ 305,750	-€ 130,732	€ 1,109,000	€ 672,518	NPV >	€ 193,078	

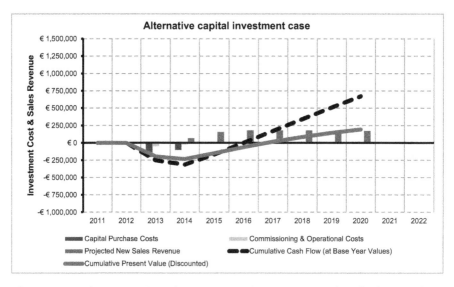

Figure 6.13 Alternative Capital Investment Case – Discounted Cash Flow Analysis

Note: the comparison is sensitive to the Discount Rate applied. Any increase in the Discount Rate will benefit those with lower positive Cash Flows in later years, whereas decreasing the Discount Rate will benefit those with higher positive Cash Flows in the later years.

Table 6.29 Discounted Cash Flow Analysis. Impact of a Delay in Investment

	NPV Method (A)	Investment & Operating Cost Delta					Discount Rate 14%	
	Year	Capital Purchase Costs	Commissioning & Operational Costs	Projected New Sales Revenue	Projected Net Annual Cost Delta	Cumulative Cash Flow (at Base Year Values)	Present Value (Discounted)	Cumulative Present Value (Discounted)
0	2011	€ 0	€ 0	€ 0	€ 0	€ 0	€ 0	€ 0
1	2012	€ 0	€ 0	€ 0	€ 0	€ 0	€ 0	€ 0
2	2013	€ 0	€ 0	€ 0	€ 0	€ 0	€ 0	€ 0
3	2014	-€ 203,450	-€ 48,750	€ 0	-€ 252,200	-€ 252,200	-€ 170,228	-€ 170,228
4	2015	-€ 102,300	-€ 24,900	€ 66,000	-€ 61,200	-€ 313,400	-€ 36,235	-€ 206,463
5	2016		-€ 9,576	€ 155,000	€ 145,424	-€ 167,976	€ 75,529	-€ 130,934
6	2017		-€ 9,576	€ 179,000	€ 169,424	€ 1,448	€ 77,187	-€ 53,747
7	2018		-€ 9,576	€ 179,000	€ 169,424	€ 170,872	€ 67,708	€ 13,961
8	2019		-€ 9,576	€ 179,000	€ 169,424	€ 340,296	€ 59,393	€ 73,354
9	2020		-€ 9,576	€ 179,000	€ 169,424	€ 509,720	€ 52,099	€ 125,453
10	2021		-€ 9,202	€ 172,000	€ 162,798	€ 672,518	€ 43,914	€ 169,367
11	2022							
	Total	-€ 305,750	-€ 130,732	€ 1,109,000	€ 672,518	NPV >	€ 169,367	

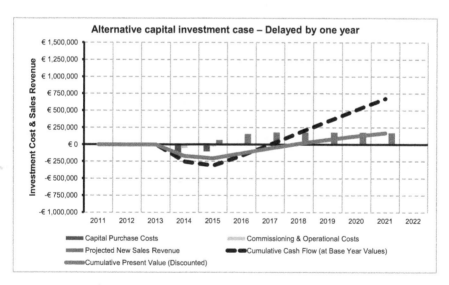

Figure 6.14 Discounted Cash Flow Analysis – Impact of a Delay in Investment

Table 6.29 and Figure 6.14 illustrates the impact of delaying the Investment opportunity by a year in relation to the current proposal in Table 6.28. From being the more attractive opportunity in comparison with the original opportunity discussed in

Table 6.27, it has now become the less attractive option. This is because we have discounted every cash flow transaction by an additional year.

Clearly, NPV analysis is sensitive to the timing of Cash Flows relative to some fixed Base Year. In cases of marginal decisions between opportunities, this might influence us to consider a more detailed phasing of the Cash Flows rather than annual totals...

Caveat augur

If more detailed Cash Flow transactions are important to us in making a true comparison between alternative investments, we can apply the principles of discounting as stated to other time periods such as quarters or months rather than years.

In these cases, however, we must substitute the annual discount rate with the respective quarterly or monthly discount rate.

For instance, for a quarterly or monthly discount rate, it is not simply a case of:

Dividing the annual discount rate by 4 or 12

Instead, we should disaggregate the compounding effect of discounting and rebuild it.

However, recognising that the Discount Rate assumed by organisations often has an element of subjective judgement associated with it (e.g. the riskiness of an investment) then an organisation may elect to use one the simple averaged value (albeit incorrect) as an approximation

---

For the Formula-philes ... Equivalent monthly and quarterly Discount Rates

If δ is the Annual Discount Rate:

$$Quarterly\ Discount\ Rate = \left(1+\delta\right)^{1/4} - 1$$

$$Monthly\ Discount\ Rate = \left(1+\delta\right)^{1/12} - 1$$

6.5.3 Internal Rate of Return (IRR)

Organisations will usually compare the Internal Rate of Return (IRR) of an investment opportunity with some internal target rate or hurdle rate, usually the Discount Rate used to generate an NPV for the level of risk associated with the investment.

Definition 6.13   Internal Rate of Return (IRR)

The Internal Rate of Return (IRR) of an investment is that Discount Rate which returns a Net Present Value (NPV) of zero, i.e. the investment breaks even over its life with no over or under recovery.

The main shortcoming of the IRR is that it disregards the time-phasing of the investment (*leading to some estimators referring to it in frustration as the 'Infernal Rate of Return'*). For instance in the examples we have just discussed in Tables 6.27 to 6.29, the IRR generated for the last two are identical:

- For Table 6.27, the IRR is 20%
- For Table 6.28, the IRR is 34.44%
- For Table 6.29, the IRR is 34.44%

There are three ways in which we can calculate the IRR:

1. In Microsoft Excel, we can use the inbuilt function **IRR(*list of values*)**. This is the easiest option if we have access to Microsoft Excel. However, there must be at least one negative value in the list of values or Excel will return an error (i.e. you cannot summate all decreasing positive values to zero)
2. We can use an iterative '*trial and error*' procedure to identify the Discount Rate that returns zero in the NPV calculation
3. We can interpolate between two discount rates, one that returns a small positive NPV, the other a small negative one. We can do this graphically as shown in Figure 6.15 giving us around 19.4%, or algebraically as depicted in Table 6.30, which gives an IRR of approximately 19.443% (compared with a value of 19.438% using the IRR function in Excel)

For the Formula-philes ... Internal Rate of Return approximation

Suppose A is the NPV returned by a Discount Rate of a%, and B is the NPV returned by a Discount Rate of b%...

Consider the curve representing the NPV for a range of Discount Rates. For any two points sufficiently close together, the curve can be approximated by a straight line.

(Continued)

The slope of the Straight Line joining the points $(a\%, A)$ and $(b\%, B)$ is the ratio of the vertical movement to the horizontal movement:

$$Slope = \frac{(B-A)}{(b-a)} \qquad (1)$$

If *IRR* is the Discount Rate which returns an NPV of zero, then if we assume that this point also lies on the line above:

$$Slope = \frac{(0-A)}{(IRR-a)} \qquad (2)$$

Eliminating Slope from (1) and (2) and inverting the relationship:

$$\frac{(IRR-a)}{(0-A)} = \frac{(b-a)}{(B-A)} \qquad (3)$$

Simplifying (3)

$$IRR = a - A\frac{(b-a)}{(B-A)}$$

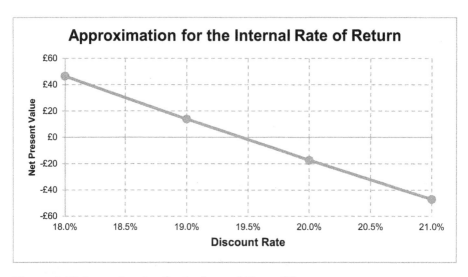

Approximation for the Internal Rate of Return

Figure 6.15 Approximation for the Internal Rate of Return

Table 6.30 Approximation to the Internal Rate of Return

Discount Rate		NPV				
a	19%	£13.83	A	$\dfrac{b-a}{B-A}$	$IRR = a - A\left(\dfrac{b-a}{B-A}\right)$	
b	20%	-£17.36	B			
b − a	1%	-£31.19	B − A	-0.0003	19.443%	

6.5.4 Payback Period

The Payback Period is an expression of how long it takes for an investment opportunity to break even, i.e. to pay back the investment.

The most basic form of Payback period is that of a Simple Payback Period in which there is no consideration of any lost alternative opportunities, i.e. the future cash flows are not discounted. (*Actually, they are discounted in terms of 'not considered' rather than discounted in the financial sense used here – just in case anyone wanted to split hairs!*)

A Modified Payback Period, however, considers how long it takes for an investment to break even from a discounted Present Value perspective.

The Payback Period (Simple or Modified) is just as important to some organisations as the overall return on the investment, especially if the company uses investment turnaround as a measure in order to reinvest in other opportunities.

Whether we are using Simple Payback or Modified Payback, we would generally calculate it from the first outlay of investment rather than the point of approval. In most cases, we can determine the Payback Period graphically. From the examples we considered earlier, we can deduce the following:

Example	Simple Payback Period	Modified Payback Period
Table 6.27	5.5 years	7.5 years
Table 6.28	3 years	4 years
Table 6.29	3 years	4 years

Note: It would be inappropriate really to talk about Payback in anything other than rounded terms especially when the Cash Flow is only considered as annual totals. Also, the higher the discount rate assumed, the longer the Modified Payback Period becomes.

6.5.5 Strengths and weaknesses of different DCF techniques

Table 6.31 summarises the strengths and weaknesses of NPV and IRR:

Table 6.31 Strengths and Weaknesses of NPV and IRR Approaches to DCF

Technique	Strength	Weakness
Net Present Value (NPV)	• Can be used to compare multiple projects over different timeframes • Can be used to compare the absolute return on investment from projects of different sizes • Variable Discount Rates can be applied in different periods if the cost of financing is expected to change • Can be used to evaluate present value where there is no cost of investment (could be free issue or charitable donation!)	• Ignores the Payback Period (breakeven point) • Sensitive to the Discount Rates used which have an element of subjectivity about them and may not be stable over a sustained period • Excel will return an NPV calculation for any range of all positive or negative values (i.e. it does not immediately highlight an investment cost or benefit that has been omitted in error)
Internal Rate of Return (IRR)	• Can be used to compare single projects against an internal benchmark return on investment • Excel return an error value when there is no investment cost	• Cannot be used to compare multiple projects over different timeframes • Ignores the size and timing of the investment – smaller or later projects may have higher IRRs but have less absolute financial impact

In general terms in comparing the use of NPV as opposed to IRR, we can conclude that the strengths of one are often a weakness of the other, so as a consequence (for the effort involved), it is recommended that both the NPV and the IRR are calculated for each investment appraisal.

6.6 Special types of formulaic normalisation techniques

There are a number of specialist formulaic techniques that can be exploited in specific circumstances. These tend to serve the purpose of normalising data, but also can be used as a predictive technique. As such they have been given their own dedicated chapters.

(*Yes, I know, I can hardly wait myself. If you can't wait either, try splashing cold water on your face and then turn to the chapter(s) of your choice.*)

- Learning Curves   see Volume IV

  In simple terms, the concept of a Learning Curve is one that recognises that the cost, effort, time and/or duration of recurring activities reduce as the cumulative quantity produced increases. If nothing else, as discussed briefly in Section 6.3.2 any data we extract from a learning curve must be normalised to reflect any difference in the cumulative quantity produced. However, there are multiple other drivers associated with learning curves that we should consider before we can say that we have '*levelled the playing field*'. So much so that we have a whole volume dedicated to the topic! (*Now, where did I put my shower cap?*)

- Norden-Rayleigh Curves   see Volume V Chapter 2

  For Design and Development projects, we may find that the relationship between cost and schedule follows a pattern of behaviour which broadly fits a Norden-Rayleigh Curve or Distribution. We will explore this as a separate topic later rather than open the discussion here, and then revisit and repeat ourselves later.

- Work Collaboration Penalties   see Volume IV Chapter 8

  In the case of large engineering projects, it is not uncommon that different elements of the design, development and production/construction activities are undertaken by a range of partners rather than by a single organisation. Here we are thinking of true legal partnership arrangements rather than mature working relationships between contractors and sub-contractors or vendors. The principle is that the more partners there are involved to share the burden of design, development and production/construction, then the greater the cost penalty to the end customer or client. This is because the downside of collaboration is that partnership arrangements are likely to increase the cost of management and integration. On the premise that these penalties are very difficult to calibrate or substantiate except by detailed and subjective audit analysis (*we don't develop products twice in order to measure the difference between the cost or effort required by a single organisation and that required by a number of organisations working in partnership*), this topic is probably more in the genre of '*food for thought*'.

6.7 Layering of normalisation for differences in analogies

The degree of normalisation we should take will depend on the difference between the context that surrounded the actual value and the context that surrounds the value we wish to estimate. One way of considering it is to draw an analogy with human relationships:

When considering something that might be classified as a Clone or an Offspring/Sibling (say, a 'repeat business' type of scenario), the level and number of normalisation dimensions to be considered should be relatively few; we may be using Primary Data directly or only considering adjustments that are date/time related, or those associated with economies of scale.

Table 6.32 Levels of Familiarity and Similarity in Estimating

Repeat or Continuing Business	**Clones**	Where the entity for which we want an estimate is exactly the same as the entity for which we have actual data, other than the date and some peripheral issues have changed. (*Hi, it's me all over again!*)
	Offspring & Siblings	Where the entity for which we want an estimate is a very close derivative of the entity for which we have actual data. (*It lives in the same house, and has a direct familial link.*)
Familiar Processes, Products or Services	**Close Cousins**	Where the entity for which we want an estimate is a more distant derivative or relative of the entity for which we have actual data. It is part of the same extended family, but there are marked differences. (*We have some things in common, and can be put together on special occasions.*)
	Near Neighbours	Where the entity for which we want an estimate is unrelated to the entity for which we have actual data, but they are familiar to us, and we have a level of understanding and many shared values. (*We live in the same place, and we have a lot in common.*)
Unfamiliar Processes, Products & Services	**Strangers**	Where the entity for which we want an estimate is very different to any entity for which we have actual data, but we recognise some of the features that characterise it. (*We're not that familiar with them, though we may have seen them before; perhaps we should ask around.*)
	Aliens	Where the entity for which we want an estimate is outside our sphere of knowledge. It is unlike anything for which we have actual data. (*We've never seen anything like them before; they might as well be from a different planet!*)

Typically, the date/time related adjustments may consider:

- Internal cost recovery rates (costing or charging rates)
- Exchange Rates
- Man-hour Performance variability (up or down)

In the case of economies of scale differences, we may need to apply a learning curve or improvement factor, but we need to ensure that we do not double-count the impact of any performance improvement that we have taken to be date/time related.

The further away from Clones or Offspring/Siblings that the analogy becomes, the more likely it is that we will have to make more complex normalisation adjustments.

For Close Cousins and Near-Neighbours we will be relying less on Primary Data, and more on Secondary Data that we have created from the Primary Source. In addition to any adjustments we would make for Clones and Offspring, our normalisation steps need to take account of differences in:

- Work Content or Scope
- Complexity
- Scale Conversion (where the comparable data is expressed in different units of measurement)

When dealing with novel or unfamiliar business (Strangers or Aliens) we are even more likely to have to consider additional and more complex normalisation activities as they potentially rely on associations to external (tertiary) data, as well as internal Primary or Secondary data. In addition to all of the above, we will have to consider the impact of changes in:

- Technology and state of readiness or maturity
- Process innovation
- Life cycle phase
- Role or mission

Each normalisation factor we apply should be done with TRACEability in mind: Transparent, Repeatable, Appropriate, Credible and Experientially-based.

In summary, the further away from our comfort blanket of tried and tested product or service we are, the more levels of adjustment we may need to consider; this is captured conceptually in Figure 6.16.

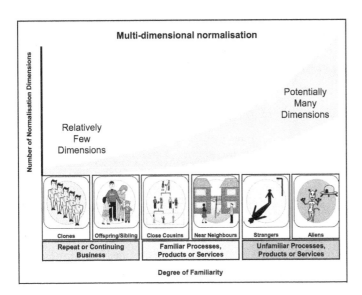

Figure 6.16 Multi-Dimensional Normalisation

6.8 Chapter review

'Oh, for a touch of normality!' was the cry.

Where does normalisation end and estimating by analogy or parametrically start? It is a moot point for which there is no easy answer. At its most basic level it can be construed that it doesn't really matter, the important point is that we have recognised that there are adjustments to be made in terms of the context of data available to us, and how we can use that sensibly to create an estimate, and that others can follow and repeat our steps – our process should be TRACEable (*Transparent, Repeatable, Appropriate, Credible and Experientially-based*).

There are relatively few basic techniques for normalising data, which can be summarised as:

a) Element Addition or Subtraction
b) Factors Rates and Ratios, or Analogical
c) Formulaic or Parametric
d) Segregation or Separation

For any estimate, the estimator may be called on to use any or all of these basic techniques depending on the number of different dimensions that need to be considered. These can be summarised in the main as those related to:

*   Correction of known demonstrable errors
*   Volume, quantity or throughput levels
*   Rule-based scale conversions (both fixed and variable rules, including dynamic or volatile rules such as currency exchange rates)
*   Differences in date or time (based on calendar and/or life cycle)
*   Key groupings defined by role or mission
*   Differences in scope (what's included or excluded)
*   Qualitative assessment of complexity or degree of difficulty to achieve

The more our subject to be estimated is unlike anything we have ever done before, the greater the number of dimensions we need to consider in terms of the context of any data we do have, and therefore we may need to normalise it in order to use for estimating.

Often, the most difficult to deal with, and the most likely to occur are those that are time-based because time will always move on. As a consequence every estimator needs to get comfortable with the concept of time travel! The definitions of 'Base Year', 'Real Year', 'Current Year', 'Constant Year' and 'Then Year' do differ within some organisations, but a useful *aide memoire* to differentiate between some of them would be:

'*Recognising the **Past** was once **Real**, we can **Then** look forward to the **Future**.*'

Economic indices are readily available from governmental sources which give credibility and consistency to those adjustments we make to take account of different rates of inflation or escalation for a range of different commodities. Strictly speaking we should always try to make time to discover the bases of the indices we use (e.g. *Laspeyres or Paasche – i.e. Fixed or Variable Shopping Bag*) so that we can interpret them correctly, although in reality it may not change our working practice in how we use them once the pressure is on! Although a somewhat tedious task, we may find that bespoke internally-generated indices are more pertinent to our own businesses than the more general market average indices.

We may get involved in comparing different investment opportunities. If we do, we will need to normalise their potential returns over a period of time. We can do this using Discounted Cash Flow techniques (DCF), which primarily consider looking at the Net Present Value or Internal Rate of Return (or preferably both) of a number of investments to see which opportunity delivers optimum benefit to the business.

'Normalisation' is a strange concept – as we said at the start, *'When was anything ever 'normal' in the life of an estimator?'*

References

Augustine, NR (1997) *Augustine's Laws (6th Edition)*, Reston, American Institute of Aeronautics and Astronautics, Inc.

Bull, A (2012) 'Paula Radcliffe steps up anti-drug focus ahead of tilt at Olympic gold', London, *The Guardian*, 13th April [online] Available from: https://www.theguardian.com/sport/2012/apr/13/paula-radcliffe-targeting-cheats-winning-gold [Accessed 11–01–2017].

Department of Defence (DoD) (2011) *Manufacturing Readiness Level (MRL) Deskbook, version 2,* May 2011.

Goodridge, P (2007) 'Methods explained: Index numbers', *Economic & Labour Market Review*, Volume 1, Number 3, March.

Havil, JG (2003) 'Exploring Euler's Constant', Princeton, Princeton University Press.

ICEAA (2009) *Cost Estimating Body of Knowledge*, Vienna, International Cost Estimating and Analysis Association.

ISO 4217:2008, *Codes for the Representation of Currencies and Funds*, International Standards Organisation.

Moore, GE (1965) 'Cramming more components onto integrated circuits', *Electronics Magazine*, Volume 38, Number 8, April 19.

NASA (2004) *NASA 2004 Cost Estimating Handbook*, Washington, NASA.

Sadin, SR, Povinelli, FP & Rosen, R (1988) '*The NASA technology push towards future space mission systems*', 39th International Astronautical Congress, Bangalore, India.

Stephenson, AG, LaPiana, LS, Mulville, DR, Rutledge, PJ, Bauer, FH, Folta, D, Dukeman, GA, Sackheim, R et al. (1999) *Mars Climate Orbiter Mishap Investigation Board Phase I Report*, Washington, NASA.

Stevenson, A & Waite, M (Eds) (2011) *Concise Oxford English Dictionary (12th Edition)*, Oxford, Oxford University Press.

Wells, HG (1895) *The Time Machine*, London, William Heinemann.

7 Pseudo-quantitative qualitative estimating techniques

If we heed the wisdom of Albert Einstein, we will recognise that sometimes we have to resort to a more qualitative or subjective assessment than rely exclusively on a quantitative or numerical evaluation. This often entails consultation with 'experts'.

Some people hate accepting inputs with the Basis of Estimate as being 'Expert Judgement' or worse 'Expert Opinion' (*implying possibly that opinion is not necessarily formed with the appropriate due judgement!*). Others see them as a necessary evil. The problem is that they do not appear to be structured, or transparent and there is sometimes an assumption that they go unchallenged. Phrases like '*Well, he/she's the expert, not me!*' and '*Their guess is better than mine; I wouldn't know where to start,*' compound the view that it would be disrespectful to challenge the experts. American Historian, Robert Dallek points out that we can and should question the wisdom of others, especially where their judgement or opinion affects our reputation (Dotinga, 2013).

Experts can be challenged to explain their thought processes. The Delphi Technique is one in which experts challenge each other.

7.1 Delphi Technique

The Delphi Technique was developed in the USA during the Cold War to predict and quantify the emerging military technologies and capabilities of the Soviet Union.

The technique relies on the premise that structured group discussion yields more realistic outcomes than individual views or opinions, especially where there is a lack of tangible evidence or knowledge to support specific quantification or evaluation. The procedure is straightforward:

1. Assemble a panel of experts with a facilitator
2. The facilitator presents the experts with the problem to be resolved
3. Each expert responds privately to the facilitator with his/her view and an outline of their rationale or reasoning
4. The facilitator shares the views of all the experts anonymously with the whole panel, and the range of potential answers
5. The experts then re-evaluate their own opinions based on the views of other experts and re-submits their current response to the facilitator
6. Steps 4 to 6 are repeated until there is no significant change in the responses received form the experts
7. When the process has converged to the 'final answer', the output is often taken to be the mean or median position (if we only want a single point deterministic answer)

The reason for anonymity is to prevent bias due to personality or hierarchy. However, in the hustle and bustle of normal business behaviour, anonymity is probably impractical and we could argue undesirable in terms of openness and honesty. A revised more practical or pragmatic procedure might be:

1. Assemble a panel of experts with a facilitator
2. The facilitator presents the experts with the problem to be resolved before the panel meets face-to-face
3. Each respond to the facilitator with his/her view and an outline of their rationale or reasoning
4. At the face-to-face meeting of the panel, the responses from each expert are presented by the facilitator
5. The experts discuss the issues and views involved
6. The experts then re-evaluate their own opinions based on the views of other experts and re-submits their current response to the facilitator
7. Steps 4 to 6 are repeated until there is no significant change in the responses received form the experts
8. When there is no further convergence in the answers output, the 'final answer' may be often taken to be the mean or median position … assuming that we only want a single point deterministic answer

However, from an estimating perspective, we have been extolling the virtues of 3-Point Range Estimates. We can use the output from the Delphi Technique directly to support this by taking the Minimum and Maximum positions as our outer bounds and an appropriate value for our Most Likely Estimate, given the following limitations:

o   In practice, it is unlikely that we will have a sufficient number of experts assembled to justify a Mode as the true Most Likely value

o   If we have only three experts, then we can take the 'middle value expert' as the Most Likely (even though it is technically the Median of a very small sample)

o   Where we have more than one other value between the Minimum and Maximum perspectives, and the mean of these remaining values is approximately halfway between the two extremes, then we can probably assume symmetry and take the halfway point as the Most Likely value

o   Where we have more than one other value between the Minimum and Maximum perspectives, but the other values are distinctly skewed to one or the other, then we can take the Most Likely value using Pearson's Rule of Thumb (Volume II Chapter 2) as:

$$\text{Mode} = 3 \times \text{Median} - 2 \times \text{Mean}$$

The use of the Delphi Technique in this way will support us in using an Ethereal Approach and Trusted Source Method without sacrificing the principles of TRACE-ability (see Chapter 3)..As well as the generation of 3-Point Estimates in this way, Delphi can be used to determine reasonable values for the probabilities of occurrence for risks and opportunities. In this case, as we will discuss in Volume V Chapter 3 Section 3.3.8, we should be using the mean values of the probabilities of occurrence that emanate from using the Delphi Technique.

In a wider sense, where we have a project that is immature, or not very well defined in terms of its scope, we can also use the Delphi Technique to develop a practical set of Assumptions, Dependencies, Opportunities, Risks and Exclusions (ADORE). By implication this can also help us define estimating options to consider.

In facilitating a Delphi Technique, we might want also to consider performing a Cross-Impact Analysis, especially if we are asking for Expert Opinion on the nature of potential impact of various factors or attributes rather than a purely quantitative response. (*That sounds like an introduction to the next section.*)

7.2 Driver Cross-Impact Analysis

In hindsight, we could have included this discussion in relation to determining the difference between Primary and Secondary Drivers in Chapter 4, but we've included it here as a complementary tool that can be used with a Delphi Technique, although it is often overlooked by estimators and other number jugglers.

A Driver Cross-Impact Analysis can be used where we have a number of competing or interacting factors or attributes that might have a bearing on the outcome of our estimating process and/or estimates that we create. It allows us to consider visually how an increase or decrease in one factor, attribute or driver increases or decreases the impact of another, and vice versa. However, it is important to note that a Cross-Impact Analysis is more than just a Correlation or Rank Correlation Matrix:

- Correlation is a two-way measure of a linear relationship between two variables
- Rank Correlation is a two-way measure of a linear OR non-linear relationship between two variables
- Neither Correlation or Rank Correlation implies Cause and Effect
- A Cross-Impact Analysis considers Cause and Effect and, therefore, implies the direction of the associated Rank Correlation (positive or negative), which is not necessarily, and is often not, Linear Correlation

Let's consider the cost to manufacture and assemble a four-wheeled motor vehicle under 3.5 tonnes, (i.e. we are excluding bikes, trikes and the likes of larger commercial goods vehicles). If we spend a few minutes thinking about different factors or characteristics that indicate or impact on cost (positively or negatively), we might come up with a list such as:

- Vehicle Kerb Weight
- Number of Seats
- Engine Size, or Engine Power
- Number of detail parts
- Number of standard features fitted and customer options available
- Proportion of 'exotic materials' used relative to 'standard materials'

> Vehicle Kerb Weight (*or Curb Weight if you prefer the American spelling*) is the weight of the vehicle, fully fuelled and with all essential hydraulic and cooling fluids etc., but excluding the driver, passenger and any other load. In other words, it is ex-Factory plus a full fuel tank.

Before we consider the Cross-Impact of these potential variables, let's firstly consider how they may be correlated with cost; this may help us to reduce the list of candidate cost drivers we have to consider.

- Weight is often a good indicator of the material content, and albeit to a lesser degree, it can be used as an indicator of work content.
- The number of seats can be an indicator of vehicle size, but not always, as car-based vans (for example) may be of a similar size as a Hatchback vehicle but have no back seats. In this case we would be comparing the cost of an empty rear end to the cost of additional seating.
- Engine Size was once an indicator of Engine Power, but with the focus on a greener economy, there has been a trend towards designing smaller engines with more power. Heavier vehicles require more power to move them in comparison with smaller vehicles.
- The number of detail parts is an interesting one (*well, I think so*) in that more parts imply more fasteners, welding or gluing etc., and therefore may imply an increase

in assembly costs. However, there is also the case that larger, more complex parts are more difficult to manufacture and therefore can incur more cost.

- The more standard features and customer options that are made available imply that the car design is more complex, with additional electrical connections with greater testing requirements. (*There's no such thing as a free lunch!*)
- 'Exotic material' is a euphemism for more expensive material than usual (otherwise automotive manufacturers would use it as standard).

On this basis, the only potential drivers that we cannot say are correlated positively with the cost to produce the vehicle is the Number of Parts and Engine Size. (Smaller, greener engines may cost more in order to recover the research investment.) Perhaps these weren't such good candidates for a cost driver; let's reject them then.

Now we can assess qualitatively (rather than quantitatively) how a change in one physical attribute is likely to impact on another attribute:

- If a change in the lead attribute causes a change in the same direction, then it has a positive impact (very similar to correlation)
- If a change in the lead attribute causes a change in the opposite direction, then we have a negative impact
- 2 signifies a strong positive impact
- 1 denotes a medium positive impact
- 0.5 is used where we have a low positive impact
- Corresponding negative values are used where we have a negative cause and effect relationship
- Zero or blank signifies little linkage between cause and effect
- A Question Mark is used to denote situations where we think that the relationship is complex in the sense that an increase in one attribute may imply an increase OR a decrease in another depending on the circumstances

It is essential that we consider this from the perspective of Cause and Effect, i.e. a change in one implies a change in the other, but not necessarily the other way around. For example, an improvement in sports results may increase spectator attendance, but an increase in spectator attendance will not necessarily improve performance on the pitch!

Table 7.1 illustrates the technique with our automotive example. For completeness, we have left the Number of Detail Parts and Engine Size as questionable Cost Drivers denoted by '?'. but have excluded them from the Driver Cross-Impact Analysis that follows:

- In terms of Kerb Weight, we have said here that this has a strong positive impact on the Engine Power required. The heavier the vehicle, the greater the power required
- Engine Power does not in itself drive changes in the other candidate variables. It is more likely to be the consequence of other changes
- The number of seats fitted will increase the Kerb Weight, but the reverse is not true

Table 7.1 Cross-Impact Analysis of Potential Automotive Cost Drivers

Physical Attribute with a Cost Impact at a Fixed Technology Maturity (Holding All Other Attributes Fixed)	Nature of Correlation with Manufacturing Cost	Cross-Impact Consequence and Direction of Change					Cause			Effect			Cross-Impact Summary			
		Kerb Weight	Engine Power	Number of Seats	No of Standard Features and Options Available	Proportion of 'Exotic Materials' Used	Sum of +ve Impacts	Sum of -ve Impacts	No of Indeterminate Impacts	Sum of +ve Consequences	Sum of -ve Consequences	No of Indeterminate Consequences	Sum of +ve or -ve Cross-Impact Cause & Effect	Total Indeterminate Cross-Impact	Net Driver Potential	Conclusion
Causative Attribute																
Kerb Weight	+ve		2				2	0	0	1.5	0	1	3.5	1	4.0	Primary Driver
Engine Power	+ve				0.5		0.5	0	0	2.5	0	1	3	1	3.5	Alternative Primary Driver
Number of Seats	+ve	1	?				1	0	1	0	0	0	1	1	1.5	Secondary Driver
No of Standard Features and Options Available	+ve	0.5	0.5				1	0	0	0.5	0	0	1.5	0	1.5	Secondary Driver
Proportion of 'Exotic Materials' Used	+ve	?					0	0	1	0	0	0	0	1	0.5	
Excluded																
Engine Size	?	<<< Excluded from Driver Cross-Impact Analysis											Average		2.2	
Number of detail parts	?															
Effect																
Sum of +ve Impacts		1.5	2.5	0	0.5	0	4.5			4.5						
Sum of -ve Impacts		0	0	0	0	0	0			0						
No of Indeterminate Impacts		1	1	0	0	0			2			2				

Key: 2 => strong positive impact; 1 => medium positive impact; 0.5 => low positive impact; Similarly for negative impacts; ? => Potentially bi-directional

- Paradoxically, the number of seats does not directly impact on the Engine Power. 2-seater sports version may have a higher performing engine than the 4/5-seater family saloon; hence the question mark
- Sometimes, exotic materials are used to improve the technical solution in terms of material stiffness or reliability, but on other occasions they may be used simply to reduce the vehicle weight, improve handling or fuel efficiency or simply from a more aesthetic perspective. This is why we have a question mark here rather than a negative value. (*Thanks to Bob Mills, formerly of Jaguar Land Rover, for putting me right on that point ... and others!*)

We can now summate the positive impacts and negative impacts, both across the columns and down the rows. The former gives a summary of the net causative impacts, and the latter gives a view of the drivers more susceptible or responsive to changes in other drivers. We can also get a view on the number of indeterminate relationships that can go either up or down. These may be the source of variation around an estimate that might be derived using the Primary Drivers alone.

To get the net result of this analysis we have taken the sum of the positive and negative impacts (ignoring signs) and subtracting half the number of indeterminate impacts. In addition, we need to be mindful of the interactions between high-scoring attributes or factors to avoid the 'double-bubble' effect. Another issue we need to consider when choosing appropriate Drivers is whether the candidate variables are 'analogue' or 'digital' in relation to the variable we are trying to estimate. In this example, Cost would be analogue, as would Kerb Weight and the proportion of Exotic Materials, whereas Engine Power and Number of Seats are more likely to be digital. Digital variables are more likely to explain step changes in the predicted variable.

However, from our Driver Cross-Impact Analysis we can conclude that:

- A high level Primary Cost Driver, e.g. vehicle Kerb Weight, is one that is not only driving the Engine Power requirement, but is also consequential to changes in multiple other attributes. As such it might be used as a high level 'independent' indicator of cost. These may be Drivers where the net score is above the average for all Drivers.
- Alternative or substitute Primary Cost Drivers may also score above the average but may in themselves be highly dependent on another higher scoring Primary Driver. In this case, the Engine Power might be a good indicator of cost, scoring almost as highly as Kerb Weight, whilst also being dependent on it. Increased Engine Power might also be one of the Customer Options offered.
- Secondary Cost Drivers might be those attributes that cannot necessarily be used as an indicator of the overall cost, but do have those 'knock-on effects' to other physical attributes. Examples here are the number of seats in the vehicle, and the number of customer options available. Typically, they would score less than the average. We could also include Engine Power as a Secondary Driver if we discount the contribution indicated by the vehicle Kerb Weight as the Primary Driver.
- Other factors appear to be more susceptible to change by other factors and as such, are probably too volatile to be used as Drivers. These would be indicated by a number of both positive and negative rank impacts down the columns. They might be used to 'tweak' (*a technical term, for 'alter slightly'*) the final Top-down estimate.

Notice how I resisted the temptation of a quip or two about Cars and Drivers? I decided to steer away from it.

7.3 A brief word or two about solution optimisation

A Driver Cross-Impact Analysis may not always be the most appropriate tool for solution optimisation, a challenge with which some estimators or cost engineers may be tasked to perform in support of their engineering colleagues. There are other tools and techniques such as Quality Function Deployment (QFD) and Analytic Hierarchy Process (AHP), that may be more appropriate. We will do no more here than raise awareness of them by outlining what they are and do.

QFD is a multi-faceted, multi-disciplinary technique that helps to interpret customer requirements or preferences (voice of the customer) into physical engineering features or attributes using Value Engineering techniques, through to product delivery (Mizuno & Akao, 1994). It is also often referred to as a 'House of Quality' as it includes a visualisation tool that resembles a house with a pitched roof with flat-roofed extensions to each side. The pitched roof of which depicts the correlation between technical requirements, depicted qualitatively rather than quantitatively to represent strong, weak, positive or negative, or no correlation. In truth, QFD is wider than just the House of Quality visualisation tool. There is an abundance of resources available on the internet on this.

Analytic Process Hierarchy was developed at the Wharton Business School in Philadelphia by Professor Thomas Saaty in the 1970s and was popularised by his book in 1980 (Saaty, 1980). AHP is more of a numerical technique that supports a structured approach to complex decision-making; however, it still requires qualitative assessments to be made but these are then recorded against a set scale from 1 to 9. It allows a goal or objective to be divided into a number of weighted priorities in which the sum of the weightings is 100%. These criteria can then be sub-divided into lower level activities. Each alternative decision outcome can then be evaluated against the goal's criteria and sub-criteria. AHP is an elegant tool but its biggest problem is that it requires us to determine eigenvectors and eigenvalues, which really requires us to have suitable specialist software to help us (*or alternatively, we would need to resort to approximations*) as these require numerical manipulation techniques that are not readily supported by Microsoft Excel unless we want to delve into VBA code.

7.4 Chapter review

In this short chapter we moved slightly left of field and looked at pseudo-quantitative qualitative techniques that might be of use to the estimator or other number juggling professional.

We concentrated on the Delphi Technique initially and how we might use that to generate a 3-Point Estimate from Subject Matter Experts using the Ethereal Approach with a Trusted Source Method. We also considered how we might use a Driver Cross-Impact Analysis to identify the potential Primary Driver from a list of candidate Drivers. These two qualitative techniques can be used in conjunction with each other.

Very briefly we introduced the concept of Quality Function Deployment and Analytic Hierarchy Process as means of optimising the development of a solution to optimise a set of customer preferences.

References

Dotinga, R (2013) 'JFK Biographer Robert Dallek looks back at the life of the 35th president', Boston, *Christian Science Monitor*, 21 November.

Mizuno, S & Akao, Y (1994) *QFD: The Customer Driven Approach to Quality Planning and Deployment*, New York, Productivity Press.

Saaty, T (1980) *The Analytic Hierarchy Process*, New York, McGraw-Hill.

8

Benford's Law as a potential measure of cost bias

If we were to ask someone to predict which leading digit occurs the most often in a large collection of numbers in a table or a database such as a detailed cost statement or project cost ledger, that person may intuitively answer '*None of them; they will all occur equally as often*', or that person may just take a random guess. In all probability, that person would be wrong. The answer is 1.

American Astronomer Stuart Newcomb (1881) was the first to document the phenomenon. He noticed that the pages containing the lower values in a set of logarithmic tables were grubbier, or more well-thumbed, than those for the higher numbers. He quantified the relationship first in what must quite literally be a 'Rule of Thumb'.

Rule of Thumb

However, it was not until Frank Benford (1938) demonstrated that this phenomenon applied also to many other naturally occurring systems of numbers, that the significance of this observation became apparent. Thereafter, this First Digit Law (or Leading Digit Law as it is sometimes called) has always been synonymous with his name ... Benford's Law.

Unlike Newton's Law of Gravity, Benford's Law is not absolute; it is a statistical observation and so the values implied by it are approximations, just as with any Probability Distribution. It applies to data that has a dimension or measurement base; it does not apply to random numbers, which are uniformly distributed.

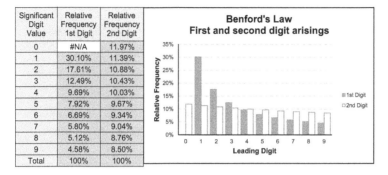

Significant Digit Value	Relative Frequency 1st Digit	Relative Frequency 2nd Digit
0	#N/A	11.97%
1	30.10%	11.39%
2	17.61%	10.88%
3	12.49%	10.43%
4	9.69%	10.03%
5	7.92%	9.67%
6	6.69%	9.34%
7	5.80%	9.04%
8	5.12%	8.76%
9	4.58%	8.50%
Total	100%	100%

Figure 8.1 Benford's Law First and Second Digit Arisings

Benford's Law extends past the original observation of Newcomb insomuch that the second and subsequent digits converge rapidly to a uniform distribution. For the second digit, we now have the possibility of a zero occurring, and the instances of zero are slightly raised in comparison with those of digit one, which are in turn slightly raised in comparison with two, three, and so on. However, these differences in the second digit frequencies are nothing like as marked as those for the first digit. Figure 8.1 gives the expected frequency that each first and second digits occurs.

It also applies to other number base systems other than 10 (such as 16 or Hexadecimal). Benford's Law does not apply to everything. It does not apply in situations where there is a bounded distribution function such as the height of adult males or females. In an over-simplification, they are more likely to occur where there is a measurement taken from zero (the lower bound) but where the largest measurement is theoretically unbounded. The lengths of rivers, the heights of tall buildings have been shown to follow Benford's Law. This suggests that there is some sense of sequential ordering that may also need to be considered. A housing estate, consisting of pre-determined house designs will have upper and lower boundaries that are determined by the physical constraints of the design and the practical response in relation to the slopes and undulations of the ground.

Newcomb (1881) and Benford (1938) showed that this natural phenomenon follows a logarithmic pattern.

For the Formula-philes: Benford's Law

Consider a collection or system of values in which the leading digit of each constituent value is an integer in range of 1 to 9.

(Continued)

Benford's Law states that the probability of the leading digit being N is given by:

$$pr(N) = \log_{10}\left(1 + \frac{1}{N}\right)$$

In a more generalised case, consider a collection or system of values expressed in number base B, in which the leading digit of each constituent value is an integer in range of 1 to B-1

Benford's Law states that the probability of the leading digit being N is given by:

$$pr(N) = \log_B\left(1 + \frac{1}{N}\right)$$

It isn't clear exactly why this phenomenon exists in so many cases where we would expect a random pattern, (many have tried). Hill (1996) is usually cited as the first to explain mathematically why this occurs, but we can get an idea of why it occurs by asking the question 'How long is a piece of string?'

For the Formula-phobes: How long is a piece of string?

The answer is, of course, 'anything we want, subject only to certain practical limits', one of which is that it must be cut. Just to be sure let's cut a piece of string three times.

However, suppose a teacher in a primary school sits the class of 33 pupils in a circle and gives each pupil a length of string cut at random from a ball. The teacher asks each pupil to cut the string they have been given into two pieces at random (*having first explained what 'random' means*). The teacher then asks each child to keep one piece of the string they've cut and to pass the other piece to the pupil on their left. The teacher now asks each pupil to cut the piece that they have been given into two random lengths. (Unless the pupils weren't listening, they should now each have three pieces of string ... the two they have now cut plus one of their original ones).

The class pupils are now asked to measure each piece of string they have. It doesn't matter whether they are doing this in Imperial or Metric measurements, so long as all the pupils are doing it the same way. For illustration here, we will assume millimetres.

The teacher then records the first digit in every one of the 99 measurements the pupils read out, giving a graph like the one shown here.

Lengths of Pieces of String

Pieces of String • Benford's Law

Let's examine the process of measurement for the longest piece of string at 788mm, cut by Johnny, the class rebel. The measurement begins at zero, passes through values of 1mm to 9mm, then through ten values beginning with the digit 1 (10mm–19mm), before it comes to a number beginning with a leading digit of 2 (20mm–29mm). The process repeats, going through successive blocks of ten like digits until it reaches 99, it then goes through a hundred leading digits of 1 before it comes to a number beginning with 2 (a hundred times) … and so on. At some stage doing that measurement process, we will reach the end of the piece of string.

During that natural process of measurement, we will have passed through more values beginning with numbers less than our measured length's leading digit, and less that are greater than it. All other pieces of string are shorter than the longest one, and so there is a higher probability that it will commence with a low value digit rather than a high value digit.

Note 1: This is not a true Benford's Law relationship but it is a reasonably close enough approximation in order to demonstrate the phenomenon.

Note 2: No children were hurt in the making of this example as safety scissors had been provided and the children supervised at all times.

Some of us may be thinking, *'That's fascinating, but other than trivia quiz games, what use is this phenomenon and the mathematical model that represents it?'*

Probably its most important use is in fraud detection. (*Honest, it's true!*) It has been used successfully to do this in many cases.

The theory and rationale for its use in detecting fraud is that when people alter their accounts fraudulently, they have a tendency to reduce selected values down and disrupt the natural pattern. Consequently, there are less values beginning with a first digit of 1 or 2, and too many values with digits 7, 8 or 9.

A word (or two) from the wise?

'Whoever is detected in a shameful fraud is ever after not believed even if they speak the truth.'

Phaedrus
Roman Poet
15 BC–50 AD

For anyone contemplating a life of crime, it isn't as easy as just rounding down everything so that all digits shift to the left. Benford's Law will also tell us the probability of values beginning with any combination of digits like 12... or 75... or 748...

8.1 Scale Invariance of Benford's Law

We might also be thinking that Benford's Law is dependent on the measurement scale we use. In other words, if we can demonstrate it to be true in the case of a Metric measurement, or a given currency then it couldn't possibly be true for the equivalent Imperial measurements or another currency, could it? After all, we will have displaced all the leading digits, either to the left or right. By shifting leading digits to the right, higher end digits like 8 or 9 trip over into the next higher order of digits beginning with 1 or 2. Similarly, by shifting leading digits to the left we have the reverse where low digit values such as 1 or 2 may trip over into lower order values with higher leading digits. Tables 8.1 and 8.2 illustrate the point with 99 pieces of string from our classroom exercise.

The left-hand side of Table 8.1 shows that our pieces of string, grouped by leading digit when measured in millimetres, follow Benford's Law. The right-hand side converts these lengths into inches using 1 inch = 25.4mm. (*In case you are wondering why the Benford values are slightly less than the percentages shown in Figure 8.1, it's because these are based on 99 pieces of string, not 100.*)

For transparency and ease of reading (*see how nice I am to you*), we have placed the right hand side of Table 8.1 as the left hand side of Table 8.2. The right hand side of this table, rearranges the measurements into leading digit groupings when measured in inches. At the foot of the table we can compare the distribution of the observed measurements in both millimetres and inches. Note that where the measurement is less than 1 inch, we ignore the leading zero and look at the first significant digit after the decimal point.

At first glance we may think that there is quite a bit of change between the two measurement scales, but if we compare the results against Benford's law graphically (Figure 8.2), then our conclusion might be that if our data obeys Benford's Law, then it will also exhibit Scale Invariance.

We can make a more rigorous assessment using a Chi-Squared Test on our data. This would show us that there is a 98.1% confidence that the first digit of our string lengths measured in millimetres is consistent with Benford's Logarithmic Distribution. When we measure them in inches, our confidence drops (*perhaps disappointingly*) to 73.07%. (*However, we did admit earlier that this 'lengths of pieces of string' exercise is not a true reflection of Benford's Law, and is only an approximation for illustration.*)

Whilst this analogy is not perfect, Scale Invariance is indeed a property of Benford's Law (Hill, 1995), and the reverse is true; Scale Invariance implies Benford's Law. We can use this as a second test of whether data is following the expected relationship by converting it to a different measurement scale. (*Statistics ... a fascinating world of accurate imprecision!*)

Table 8.1 Lengths of Pieces of String Measured in Millimetres and Inches (1)

Pieces of string measured in millimetres

Count	First Digit Groupings								
	1	2	3	4	5	6	7	8	9
1	192	287	396	471	594	641	716	89	99
2	179	283	393	449	591	68	75	88	97
3	175	272	386	442	538	63	74	87	96
4	174	238	381	434	528	62	72		94
5	169	236	363	418	501	61	71		
6	167	229	352	417	58				
7	165	226	319	413	55				
8	162	221	316	48	52				
9	161	215	313	44	51				
10	139	214	39						
11	134	208	37						
12	128	29	35						
13	125	27	33						
14	119	26							
15	117	25							
16	116	24							
17	114	22							
18	111	21							
19	109								
20	108								
21	107								
22	106								
23	103								
24	101								
25	19								
26	18								
27	17								
28	15								
29	14								
30	13								
31	12								
32	11								
33	10								
Total	33	18	13	9	9	5	5	3	4
Benford	29.8	17.4	12.4	9.6	7.8	6.6	5.7	5.1	4.5

Pieces of string measured in inches

Count	First Digit Groupings (When Measured in Millimetres)								
	1	2	3	4	5	6	7	8	9
1	7.56	11.30	15.59	18.54	23.39	25.24	28.19	3.50	3.90
2	7.05	11.14	15.47	17.68	23.27	2.68	2.95	3.46	3.82
3	6.89	10.71	15.20	17.40	21.18	2.48	2.91	3.43	3.78
4	6.85	9.37	15.00	17.09	20.71	2.44	2.83		3.70
5	6.65	9.29	14.29	16.46	19.72	2.40	2.80		
6	6.57	9.02	13.86	16.42	2.28				
7	6.50	8.90	12.56	16.26	2.17				
8	6.38	8.70	12.44	1.89	2.05				
9	6.34	8.46	12.32	1.73	2.01				
10	5.47	8.43	1.54						
11	5.28	8.19	1.46						
12	5.04	1.14	1.38						
13	4.92	1.06	1.30						
14	4.69	1.02							
15	4.61	0.98							
16	4.57	0.94							
17	4.49	0.87							
18	4.37	0.83							
19	4.29								
20	4.25								
21	4.21								
22	4.17								
23	4.06								
24	3.98								
25	0.75								
26	0.71								
27	0.67								
28	0.59								
29	0.55								
30	0.51								
31	0.47								
32	0.43								
33	0.39								
Total	33	18	13	9	9	5	5	3	4
Benford	29.8	17.4	12.4	9.6	7.8	6.6	5.7	5.1	4.5

Table 8.2 Lengths of Pieces of String Measured in Millimetres and Inches (2)

Pieces of string measured in inches

Count	First Digit Groupings (When Measured in Millimetres)								
	1	2	3	4	5	6	7	8	9
1	7.56	11.30	15.59	18.54	23.39	25.24	28.19	3.50	3.90
2	7.05	11.14	15.47	17.68	23.27	2.68	2.95	3.46	3.82
3	6.89	10.71	15.20	17.40	21.18	2.48	2.91	3.43	3.78
4	6.85	9.37	15.00	17.09	20.71	2.44	2.83		3.70
5	6.65	9.29	14.29	16.46	19.72	2.40	2.80		
6	6.57	9.02	13.86	16.42	2.28				
7	6.50	8.90	12.56	16.26	2.17				
8	6.38	8.70	12.44	1.89	2.05				
9	6.34	8.46	12.32	1.73	2.01				
10	5.47	8.43	1.54						
11	5.28	8.19	1.46						
12	5.04	1.14	1.38						
13	4.92	1.06	1.30						
14	4.69	1.02							
15	4.61	0.98							
16	4.57	0.94							
17	4.49	0.87							
18	4.37	0.83							
19	4.29								
20	4.25								
21	4.21								
22	4.17								
23	4.06								
24	3.98								
25	0.75								
26	0.71								
27	0.67								
28	0.59								
29	0.55								
30	0.51								
31	0.47								
32	0.43								
33	0.39								
Total	33	18	13	9	9	5	5	3	4
Benford	29.8	17.4	12.4	9.6	7.8	6.6	5.7	5.1	4.5

Pieces of string measured in inches

Count	First Digit Groupings								
	1	2	3	4	5	6	7	8	9
1	11.30	23.39	3.98	4.92	5.47	6.89	7.56	9.02	9.37
2	11.14	23.27	0.39	4.69	5.28	6.85	7.05	8.90	9.29
3	10.71	21.18	3.50	4.61	5.04	6.65	0.75	8.70	0.98
4	1.14	20.71	3.46	4.57	0.59	6.57	0.71	8.46	0.94
5	1.06	2.28	3.43	4.49	0.55	6.50		8.43	
6	1.02	2.17	3.90	4.37	0.51	6.38		8.19	
7	15.59	2.05	3.82	4.29		6.34		0.87	
8	15.47	2.01	3.78	4.25		0.67		0.83	
9	15.20	25.24	3.70	4.21					
10	15.00	2.68		4.17					
11	14.29	2.48		4.06					
12	13.86	2.44		0.47					
13	12.56	2.40		0.43					
14	12.44	28.19							
15	12.32	2.95							
16	1.54	2.91							
17	1.46	2.83							
18	1.38	2.80							
19	1.30								
20	18.54								
21	17.68								
22	17.40								
23	17.09								
24	16.46								
25	16.42								
26	16.26								
27	1.89								
28	1.73								
29	19.72								
30									
31									
32									
33									
Total	29	18	9	13	6	8	4	8	4
Benford	29.8	17.4	12.4	9.6	7.8	6.6	5.7	5.1	4.5

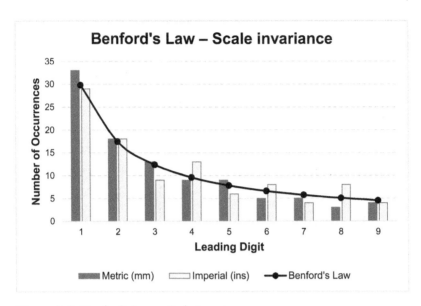

Figure 8.2 Benford's Law – Scale Invariance

8.2 Potential use of Benford's Law in estimating

So, perhaps you are now thinking 'How can this property be used by estimators and other number jugglers?' Well, our thinking might be along the lines of …

- If they use Benford's Law for Fraud detection, then we might expect that our recorded costs against a detailed Work Breakdown Structure should also follow Benford's Law, and the initial budget or estimate should likewise follow that same law.
- If that is true, then surely it follows that if there are too high a proportion of WBS Element values beginning with the higher digits, then perhaps that is an indicator that the estimate or budget is too optimistic.
- On the other hand, too great a proportion of WBS Elements with ones and/or twos may be an indication that the estimate or budget is somewhat pessimistic (*some people may prefer to use the technical term 'stuffed'!*).

However, sorry to disappoint you but this is unlikely to work … so, before we all go out and check the latest budgets and start accusing the Project Manager of fiddling the budget, let's just take a step back and think again.

A WBS is a convenient means or summarising work and costs for management. Benford's Law is a phenomenon that works best at the transactional level. If we have a sufficiently detailed Bottom-up Estimate with a number of 'Trusted Source' inputs, then it may be appropriate to use Benford's Law to identify potential positive or negative bias in the constituent elements of the Bottom-up Estimate. However, wherever we have a man-made summarisation process, or we use a Top-down Approach to estimating, this natural order of things at the transactional level begins to break down. We can demonstrate this by looking at a Monte Carlo Simulation on the Baseline Task (excluding Risks or Opportunities). Our inputs will almost always be based on known or assumed Probability Distributions, and therefore, we will have constrained both the lower and upper end of the potential range to a large degree, thus breaking one of the requirements for Benford's Law to be relevant.

So in what circumstances, does Benford's Law apply? The following characteristics are suggested:

- Numeric data that was a dimension or measurement scale
- Generated randomly from different distributions (Hill, 1995)
- Data that is not restricted by maxima or minima (although by implication most measurements begin from zero, and most data has a natural maximum)
- Large sets of data (which is not untypical of any statistical distribution)
- The values cover several orders of magnitude (e.g. data values can theoretically cover units, tens, hundreds, thousands, etc.)

Let's look at an example where Benford's Law might be used. Suppose we want to test whether a number of quotations received from a range of vendors follow Benford's Law. We can count the number of occurrences that each first digit occurs and compare these with Benford's Law. Now we will always get a variation in the specific results (*that's the joy of statistics, nothing is ever exact and why the world needs estimators*) so we need to perform the relevant statistical test. The Chi-Squared Test (see Volume II Chapter 6) allows us to compare an observed distribution with a theoretical or expected distribution.

In this example in Figure 8.3, using the Chi-Squared Test, as noted above, there is only a 38% probability that the vendor quotations match Benford's Law. We might want to investigate whether this could be down to:

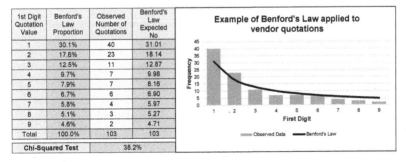

Figure 8.3 Example of Benford's Law Applied to Vendor Quotations

- Some vendors rounding up values; the difficulty is finding out which ones. If everyone did it then it would be masked by the Scale Invariance Property
- A consequence of our procurement policy. For example, there may be a policy to place orders with preferred suppliers who may operate a policy of Minimum Order Values
- A mixture of the two

In contrast, if we had too many higher order values, this may be indicative of vendors submitting 'loss leaders' whereby there is a reasonable chance that they can recover this on change orders, follow-on orders, spares etc.

8.3 Chapter review

We began by considering the phenomenon that is Benford's Law that states that under certain (fairly loose) conditions, the leading digits of apparently random numbers are more likely to begin with a 1. In fact, each subsequent number from 2 through to 9 is less likely to appear as the leading digit than the preceding lower value digit. This pattern can be extended to leading pairs of digits also.

One essential property of Benford's Law is that it is Scale Invariant in that if it is shown to work on a set of data using one measurement scale, then it will also work for another measurement scale.

This phenomenon is frequently used in fraud detection, but it is offered here for consideration to detect the presence of estimator or vendor bias in major bid tenders and quotations.

References

Benford, F (1938) 'The law of anomalous numbers', *Proc. Am. Philos. Soc.* Volume 78, Number 4: pp.551–72.

Hill, TP (1995) 'Base-invariance implies Benford's law', Proc. Amer. Math. Soc. Volume 123, Number 3: pp.887–95.

Hill, TP (1996) 'A statistical derivation of the Significant-Digit Law', *Statistical Science,* Volume 10: pp.354–63.

Newcomb, S (1881) 'Note on the frequency of use of the different digits in natural numbers', *American Journal of Mathematics,* Volume 4, Number 1: pp.39–40.

Glossary of estimating and forecasting terms

This Glossary reflects those Estimating Terms that are either in common usage or have been defined for the purposes of this series of guides. Not all the terms are used in every volume, but where they do occur, their meaning is intended to be consistent.

3-Point Estimate   A 3-Point Estimate is an expression of uncertainty around an Estimate Value. It usually expresses Optimistic, Most Likely and Pessimistic Values.

Accuracy   Accuracy is an expression of how close a measurement, statistic or estimate is to the true value, or to a defined standard.

Actual Cost (AC)   See Earned Value Management Abbreviations and Terminology

ACWP (Actual Cost of Work Performed) or Actual Cost (AC)   See Earned Value Management Terminology

Additive/Subtractive Time Series Model   See Time Series Analysis

Adjusted R-Square   Adjusted R-Square is a measure of the "Goodness of Fit" of a Multi-Linear Regression model to a set of data points, which reduces the Coefficient of Determination by a proportion of the Unexplained Variance relative to the Degrees of Freedom in the model, divided by the Degrees of Freedom in the Sum of Squares Error.

ADORE (Assumptions, Dependencies, Opportunities, Risks, Exclusions)   See Individual Terms.

Alternative Hypothesis   An Alternative Hypothesis is that supposition that the difference between an observed value and another observed or assumed value or effect, cannot be legitimately attributable to random sampling or experimental error. It is usually denoted as H_1.

Analogous Estimating Method or Analogy   See Analogical Estimating Method.

Analogical Estimating Method   The method of estimating by Analogy is a means of creating an estimate by comparing the similarities and/or differences between two things, one of which is used as the reference point against which rational adjustments for differences between the two things are made in order establish an estimate for the other.

Approach   See Estimating Approach.

Arithmetic Mean or Average   The Arithmetic Mean or Average of a set of numerical data values is a statistic calculated by summating the values of the individual terms and dividing by the number of terms in the set.

Assumption   An Assumption is something that we take to be broadly true or expect to come to fruition in the context of the Estimate.

Asymptote   An Asymptote to a given curve is a straight line that tends continually closer in value to that of the curve as they tend towards infinity (positive or negative). The difference between the asymptote and its curve reduces towards but never reaches zero at any finite value.

AT (Actual Time)   See Earned Value Management Abbreviations and Terminology.

Average   See Arithmetic Mean.

Average (Mean) Absolute Deviation (AAD)   The Mean or Average Absolute Deviation of a range of data is the average 'absolute' distance of each data point from the Arithmetic Mean of all the data points, ignoring the sign depicting whether each point is less than or greater than the Arithmetic Mean.

Axiom   An Axiom is a statement or proposition that requires no proof, being generally accepted as being self-evidently true at all times.

BAC (Budget At Completion)   See Earned Value Management Abbreviations and Terminology.

Base Year Values   'Base Year Values' are values that have been adjusted to be expressed relative to a fixed year as a point of reference e.g., for contractual price agreement.

Basis of Estimate (BoE)   A Basis of Estimate is a series of statements that define the assumptions, dependencies and exclusions that bound the scope and validity of an estimate. A good BoE also defines the approach, method and potentially techniques used, as well as the source and value of key input variables, and as such supports Estimate TRACEability.

BCWP (Budgeted Cost of Work Performed)   See Earned Value Management Abbreviations and Terminology.

BCWS (Budgeted Cost of Work Scheduled)   See Earned Value Management Abbreviations and Terminology.

Benford's Law   Benford's Law is an empirical observation that in many situations the first or leading digit in a set of apparently random measurements follows a repeating pattern that can be predicted as the Logarithm of one plus the reciprocal of the leading digit. It is used predominately in the detection of fraud.

Bessel's Correction Factor   In general, the variance (and standard deviation) of a data sample will understate the variance (and standard deviation) of the underlying data population. Bessel's Correction Factor allows for an adjustment to be made so that the sample variance can be used as an unbiased estimator of the population variance. The adjustment requires that the Sum of Squares of the Deviations from the Sample Mean be divided one less than the number of observations or data points i.e. n-1 rather than the more intuitive the number of observations. Microsoft Excel takes this adjustment into account.

Bottom-up Approach   In a Bottom-up Approach to estimating, the estimator identifies the lowest level at which it is appropriate to create a range of estimates based on the task definition available, or that can be inferred. The overall estimate, or higher level summaries, typically through a Work Breakdown Structure, can be produced through incremental aggregation of the lower level estimates. A Bottom-up Approach requires a good definition of the task to be estimated, and is frequently referred to as detailed estimating or as engineering build-up.

Chauvenet's Criterion   A test for a single Outlier based on the deviation Z-Score of the suspect data point.

Chi-Squared Test or χ^2-Test The Chi-Squared Test is a "goodness of fit" test that compares the variance of a sample against the variance of a theoretical or assumed distribution.

Classical Decomposition Method (Time Series) Classical Decomposition Method is a means of analysing data for which there is a seasonal and/or cyclical pattern of variation. Typically, the underlying trend is identified, from which the average deviation or variation by season can be determined. The method can be used for multiplicative and additive/subtractive Time Series Models.

Closed Interval A Closed Continuous Interval is one which includes its endpoints, and is usually depicted with square brackets: [Minimum, Maximum].

Coefficient of Determination The Coefficient of Determination is a statistical index which measures how much of the total variance in one variable can be explained by the variance in the other variable. It provides a measure of how well the relationship between two variables can be represented by a straight line.

Coefficient of Variation (CV) The Coefficient of Variation of a set of sample data values is a dimensionless statistic which expresses the ratio of the sample's Standard Deviation to its Arithmetic Mean. In the rare cases where the set of data is the entire population, then the Coefficient of Variation is expressed as the ratio of the population's Standard Deviation to its Arithmetic Mean. It can be expressed as either a decimal or percentage.

Collaborative Working Collaborative Working is a term that refers to the management strategy of dividing a task between multiple partners working towards a common goal where there a project may be unviable for a single organisation. There is usually a cost penalty of such collaboration as it tends to create duplication in management and in integration activities.

Collinearity & Multicollinearity Collinearity is an expression of the degree to which two supposedly independent predicator variables are correlated in the context of the observed values being used to model their relationship with the dependent variable that we wish to estimate. Multicollinearity is an expression to which collinearity can be observed across several predicator variables.

Complementary Cumulative Distribution Function (CCDF) The Complementary Cumulative Distribution Function is the theoretical or observed probability of that variable being greater than a given value. It is calculated as the difference between 1 (or 100%) and the Cumulative Distribution Function, 1-CDF.

Composite Index A Composite Index is one that has been created as the weighted average of a number of other distinct Indices for different commodities.

Concave Curve A curve in which the direction of curvature appears to bend towards a viewpoint on the x-axis, similar to one that would be observed when viewing the inside of a circle or sphere.

Cone of Uncertainty A generic term that refers to the empirical observation that the range of estimate uncertainty or accuracy improves through the life of a project. It is typified by its cone or funnel shape appearance.

Confidence Interval A Confidence Interval is an expression of the percentage probability that data will lie between two distinct Confidence Levels, known as the Lower and Upper Confidence Limits, based on a known or assumed distribution of data from either a sample or an entire population.
See also Prediction Interval.

Confidence Level A Confidence Level is an expression of the percentage probability that data selected at random from a known or assumed distribution of data (either a sample or an entire population), will be less than or equal to a particular value.

Confidence Limits   The Lower and Upper Confidence Limits are the respective Confidence Levels that bound a Confidence Interval, and are expressions of the two percentage probabilities that data will be less or equal to the values specified based on the known or assumed distribution of data in question from either a sample or an entire population. See also Confidence Interval.

Constant Year Values   'Constant Year Values' are values that have been adjusted to take account of historical or future inflationary effects or other changes, and are expressed in relation to the Current Year Values for any defined year. They are often referred to as 'Real Year Values'.

Continuous Probability Distribution   A mathematical expression of the relative theoretical probability of a random variable which can take on any value from a real number range. The range may be bounded or unbounded in either direction.

Convex Curve   A curve in which the direction of curvature appears to bend away from a viewpoint on the x-axis, similar to one that would be observed when viewing the outside of a circle or sphere.

Copula   A Copula is a Multivariate Probability Distribution based exclusively on a number Uniform Marginal Probability Distributions (one for each variable).

Correlation   Correlation is a statistical relationship in which the values of two or more variables exhibit a tendency to change in relationship with one other. These variables are said to be positively (or directly) correlated if the values tend to move in the same direction, and negatively (or inversely) correlated if they tend to move in opposite directions.

Cost Driver   See Estimate Drivers.

Covariance   The Covariance between a set of paired values is a measure of the extent to which the paired data values are scattered around the paired Arithmetic Means. It is the average of the product of each paired variable from its Arithmetic Mean.

CPI (Cost Performance Index)   See Earned Value Management Abbreviations and Terminology.

Crawford's Unit Learning Curve   A Crawford Unit Learning Curve is an empirical relationship that expresses the reduction in time or cost of each unit produced as a power function of the cumulative number units produced.

Critical Path   The Critical Path at a point in time depicts the string of dependent activities or tasks in a schedule for which there is no float or queuing time. As such the length of the Critical Path represents the quickest time that the schedule can be currently completed based on the current assumed activity durations.

Cross-Impact Analysis   A Cross-Impact Analysis is a qualitative technique used to identify the most significant variables in a system by considering the impact of each variable on the other variables.

Cumulative Average   A Point Cumulative Average is a single term value calculated as the average of the current and all previous consecutive recorded input values that have occurred in a natural sequence.

A Moving Cumulative Average, sometimes referred to as a Cumulative Moving Average, is an array (a series or range of ordered values) of successive Point Cumulative Average terms calculated from all previous consecutive recorded input values that have occurred in a natural sequence.

Cumulative Distribution Function (CDF)   The Cumulative Distribution Function of a Discrete Random Variable expresses the theoretical or observed probability of that variable being less than or equal to any given value. It equates to the sum of the probabilities of achieving that value and each successive lower value.

The Cumulative Distribution Function of a Continuous Random Variable expresses the theoretical or observed probability of that variable being less than or equal to any given value. It equates to the area under the Probability Density Function curve to the left of the value in question.

See also the Complementary Cumulative Distribution Function.

Current Year (or Nominal Year) Values 'Current Year Values' are historical values expressed in terms of those that were current at the historical time at which they were incurred. In some cases, these may be referred to as 'Nominal Year Values'.

CV (Cost Variance) See Earned Value Management Abbreviations and Terminology

Data Type Primary Data is that which has been taken directly from its source, either directly or indirectly, without any adjustment to its values or context.

Secondary Data is that which has been taken from a known source, but has been subjected to some form of adjustment to its values or context, the general nature of which is known and has been considered to be appropriate.

Tertiary Data is data of unknown provenance. The specific source of data and its context is unknown, and it is likely that one or more adjustments of an unknown nature have been made, in order to make it suitable for public distribution.

Data Normalisation Data Normalisation is the act of making adjustments to, or categorisations of, data to achieve a state where data the can be used for comparative purposes in estimating.

Decile A Decile is one of ten subsets from a set of ordered values which nominally contain a tenth of the total number of values in each subset. The term can also be used to express the values that divide the ordered values into the ten ordered subsets.

Degrees of Freedom Degrees of Freedom are the number of different factors in a system or calculation of a statistic that can vary independently.

DeJong Unit Learning Curve A DeJong Unit Learning Curve is a variation of the Crawford Unit Learning Curve that allows for an incompressible or 'unlearnable' element of the task, expressed as a fixed cost or time.

Delphi Technique The Delphi Technique is a qualitative technique that promotes consensus or convergence of opinions to be achieved between diverse subject matter experts in the absence of a clear definition of a task or a lack of tangible evidence.

Dependency A Dependency is something to which an estimate is tied, usually an uncertain event outside of our control or influence, which if it were not to occur, would potentially render the estimated value invalid. If it is an internal dependency, the estimate and schedule should reflect this relationship

Descriptive Statistic A Descriptive Statistic is one which reports an indisputable and repeatable fact, based on the population or sample in question, and the nature of which is described in the name of the Statistic.

Discount Rate The Discount Rate is the percentage reduction used to calculate the present-day values of future cash flows. The discount rate often either reflects the comparable market return on investment of opportunities with similar levels of risk, or reflects an organisation's Weighted Average Cost of Capital (WACC), which is based on the weighted average of interest rates paid on debt (loans) and shareholders' return on equity investment.

Discounted Cash Flow (DCF) Discounted Cash Flow (DCF) is a technique for converting estimated or actual expenditures and revenues to economically comparable values at a common point in time by discounting future cash flows by an agreed percentage discount rate per time period, based on the cost to the organisation of borrowing money, or the average return on comparable investments.

Discrete Probability Distribution   A mathematical expression of the theoretical or empirical probability of a random variable which can only take on predefined values from a finite range.

Dixon's Q-Test   A test for a single Outlier based on the distance between the suspect data point and its nearest neighbour in comparison with the overall range of the data.

Driver   See Estimate Drivers.

Earned Value (EV)   See Earned Value Management Terminology.

Earned Value Management (EVM)   Earned Value Management is a collective term for the management and control of project scope, schedule and cost.

Earned Value Analysis   Earned Value Analysis is a collective term used to refer to the analysis of data gathered and used in an Earned Value Management environment.

Earned Value Management Abbreviations and Terminology (Selected terms only)

   ACWP (Actual Cost of Work Performed) sometimes referred to as Actual Cost (AC)   Each point represents the cumulative actual cost of the work completed or in progress at that point in time. The curve represents the profile by which the actual cost has been expended for the value achieved over time.

   AT (Actual Time)   AT measures the time from start to time now.

   BAC (Budget At Completion)   The BAC refers to the agreed target value for the current scope of work, against which overall performance will be assessed.

   BCWP (Budget Cost of Work Performed) sometimes referred to as Earned Value (EV)   Each point represents the cumulative budgeted cost of the work completed or in progress to that point in time. The curve represents the profile by which the budgeted cost has been expended over time. The BCWP is expressed in relation to the BAC (Budget At Completion).

   BCWS sometimes referred to as Planned Value (PV)   Each point represents the cumulative budgeted cost of the work planned to be completed or to be in progress to that point in time. The curve represents the profile by which the budgeted cost was planned to be expended over time. The BCWS is expressed in relation to the BAC (Budget At Completion).

   CPI (Cost Performance Index)   The CPI is an expression of the relative performance from a cost perspective and is the ratio of Earned Value to Actual Cost (EV/AC) or (BCWP/ACWP).

   CV (Cost Variance)   CV is a measure of the cumulative Cost Variance as the difference between the Earned Value and the Actual Cost (EV − AC) or (BCWP − ACWP).

   ES (Earned Schedule)   ES measures the planned time allowed to reach the point that we have currently achieved.

   EAC (Estimate At Completion) sometimes referred to as FAC (Forecast At Completion)   The EAC or FAC is the sum of the actual cost to date for the work achieved, plus an estimate of the cost to complete any outstanding or incomplete activity or task in the defined scope of work

   ETC (Estimate To Completion)   The ETC is an estimate of the cost that is likely to be expended on the remaining tasks to complete the current scope of agreed work. It is the difference between the Estimate At Completion and the current Actual Cost (EAC − ACWP or AC).

SPI (Schedule Performance Index)   The SPI is an expression of the relative schedule performance expressed from a cost perspective and is the ratio of Earned Value to Planned Value (EV/PV) or (BCWP/BCWS). It is now considered to be an inferior measure of true schedule variance in comparison with SPI(t).

SPI(t)   The SPI(t) is an expression of the relative schedule performance and is the ratio of Earned Schedule to Actual Time (ES/AT).

SV (Schedule Variance)   SV is a measure of the cumulative Schedule Variance measured from a Cost Variance perspective, and is the difference between the Earned Value and the Planned Value (EV − PV) or (BCWP − BCWS). It is now considered to be an inferior measure of true schedule variance in comparison with SV(t).

SV(t)   SV(t) is a measure of the cumulative Schedule Variance and is the difference between the Earned Schedule and the Actual Time (ES − AT).

Equivalent Unit Learning   Equivalent Unit Learning is a technique that can be applied to complex programmes of recurring activities to take account of Work-in-Progress and can be used to give an early warning indicator of potential learning curve breakpoints. It can be used to supplement traditional completed Unit Learning Curve monitoring.

ES (Earned Schedule)   See Earned Value Management Abbreviations and Terminology

Estimate   An Estimate for 'something' is a numerical expression of the approximate value that might reasonably be expected to occur based on a given context, which is described and is bounded by a number of parameters and assumptions, all of which are pertinent to and necessarily accompany the numerical value provided.

Estimate At Completion (EAC) and **Estimate To Completion (ETC)**   See Earned Value Management Abbreviations and Terminology.

Estimate Drivers   A Primary Driver is a technical, physical, programmatic or transactional characteristic that either causes a major change in the value being estimated or in a major constituent element of it, or whose value itself changes correspondingly with the value being estimated, and therefore, can be used as an indicator of a change in that value.

A Secondary Driver is a technical, physical, programmatic or transactional characteristic that either causes a minor change in the value being estimated or in a constituent element of it, or whose value itself changes correspondingly with the value being estimated and can be used as an indicator of a subtle change in that value.

Cost Drivers are specific Estimate Drivers that relate to an indication of Cost behaviour.

Estimate Maturity Assessment (EMA)   An Estimate Maturity Assessment provides a 'health warning' on the maturity of an estimate based on its Basis of Estimate, and takes account of the level of task definition available and historical evidence used.

Estimating Approach   An Estimating Approach describes the direction by which the lowest level of detail to be estimated is determined.
See also Bottom-up Approach, Top-down Approach and Ethereal Approach.

Estimating Method   An Estimating Method is a systematic means of creating an estimate, or an element of an estimate. An Estimating Methodology is a set or system of Estimating Methods.
See also Analogous Method, Parametric Method and Trusted Source Method.

Estimating Metric   An Estimating Metric is a value or statistic that expresses a numerical relationship between a value for which an estimate is required, and a Primary or Secondary Driver (or parameter) of that value, or in relation to some fixed reference point.
See also Factor, Rate and Ratio.

Estimating Procedure   An Estimating Procedure is a series of steps conducted in a certain manner and sequence to optimise the output of an Estimating Approach, Method and/ or Technique.

Estimating Process   An Estimating Process is a series of mandatory or possibly optional actions or steps taken within an organisation, usually in a defined sequence or order, in order to plan, generate and approve an estimate for a specific business purpose.

Estimating Technique   An Estimating Technique is a series of actions or steps conducted in an efficient manner to achieve a specific purpose as part of a wider Estimating Method. Techniques can be qualitative as well as quantitative.

Ethereal Approach   An Ethereal Approach to Estimating is one in which values are accepted into the estimating process, the provenance of which is unknown and at best may be assumed. These are values often created by an external source for low value elements of work, or by other organisations with acknowledged expertise. Other values may be generated by Subject Matter Experts internal to the organisation where there is insufficient definition or data to produce an estimate by a more analytical approach.

The Ethereal Approach should be considered the approach of last resort where low maturity is considered acceptable. The approach should be reserved for low value elements or work, and situations where a robust estimate is not considered critical.

Excess Kurtosis   The Excess Kurtosis is an expression of the relative degree of Peakedness or flatness of a set of data values, relative to a Normal Distribution. Flatter distributions with a negative Excess Kurtosis are referred to as Platykurtic; Peakier distributions with a positive Excess Kurtosis are termed Leptokurtic; whereas those similar to a Normal Distribution are said to be Mesokurtic. The measure is based on the fourth power of the deviation around the Arithmetic Mean.

Exclusion   An Exclusion is condition or set of circumstances that have been designated to be out of scope of the current estimating activities and their output.

Exponential Function   An Exponential Function of two variables is one in which the Logarithm of the dependent variable on the vertical axis produces a monotonic increasing or decreasing Straight Line when plotted against the independent variable on the horizontal axis.

Exponential Smoothing   Exponential Smoothing is a 'single-point' predictive technique which generates a forecast for any period based on the forecast made for the prior period, adjusted for the error in that prior period's forecast.

Extrapolation   The act of estimating a value extrinsic to or outside the range of the data being used to determine that value. See also Interpolation.

Factored or Expected Value Technique   A technique that expresses an estimate based on the weighted sum of all possible values multiplied by the probability of arising.

Factors, Rates and Ratios   See individual terms: Factor Metric, Rate Metric and Ratio Metric

Factor Metric   A Factor is an Estimating Metric used to express one variable's value as a percentage of another variable's value.

F-Test   The F-Test is a "goodness of fit" test that returns the cumulative probability of getting an F-Statistic less than or equal to the ratio inferred by the variances in two samples.

Generalised Exponential Function   A variation to the standard Exponential Function which allows for a constant value to exist in the dependent or predicted variable's value. It effectively creates a vertical shift in comparison with a standard Exponential Function.

Generalised Extreme Studentised Deviate   A test for multiple Outliers based on the deviation Z-Score of the suspect data point.

Generalised Logarithmic Function A variation to the standard Logarithmic Function which allows for a constant value to exist in the independent or predictor variable's value. It effectively creates a horizontal shift in comparison with a standard Logarithmic Function.

Generalised Power Function A variation to the standard Power Function which allows for a constant value to exist in either or both the independent and dependent variables' value. It effectively creates a horizontal and/or vertical shift in comparison with a standard Power Function.

Geometric Mean The Geometric Mean of a set of n numerical data values is a statistic calculated by taking the nth root of the product of the n terms in the set.

Good Practice Spreadsheet Modelling (GPSM) Good Practice Spreadsheet Modelling Principles relate to those recommended practices that should be considered when developing a Spreadsheet in order to help maintain its integrity and reduce the risk of current and future errors.

Grubbs' Test A test for a single Outlier based on the deviation Z-Score of the suspect data point.

Harmonic Mean The Harmonic Mean of a set of n numerical data values is a statistic calculated by taking the reciprocal of the Arithmetic Mean of the reciprocals of the n terms in the set.

Heteroscedasticity Data is said to exhibit Heteroscedasticity if data variances are not equal for all data values.

Homoscedasticity Data is said to exhibit Homoscedasticity if data variances are equal for all data values.

Iglewicz and Hoaglin's M-Score (Modified Z-Score) A test for a single Outlier based on the Median Absolute Deviation of the suspect data point.

Index An index is an empirical average factor used to increase or decrease a known reference value to take account of cumulative changes in the environment, or observed circumstances, over a period of time. Indices are often used as to normalise data.

Inferential Statistic An Inferential Statistic is one which infers something, often about the wider data population, based on one or more Descriptive Statistics for a sample, and as such, it is open to interpretation … and disagreement.

Inherent Risk in Spreadsheets (IRiS) IRiS is a qualitative assessment tool that can be used to assess the inherent risk in spreadsheets by not following Good Practice Spreadsheets Principles.

Interdecile Range The Interdecile Range comprises the middle eight Decile ranges and represents the 80% Confidence Interval between the 10% and 90% Confidence Levels for the data.

Internal Rate of Return The Internal Rate of Return (IRR) of an investment is that Discount Rate which returns a Net Present Value (NPV) of zero, i.e. the investment breaks even over its life with no over or under recovery.

Interpolation The act of estimating an intermediary or intrinsic value within the range of the data being used to determine that value. See also Extrapolation.

Interquantile Range An Interquantile Range is a generic term for the group of Quantiles that form a symmetrical Confidence Interval around the Median by excluding the first and last Quantile ranges.

Interquartile Range The Interquartile Range comprises the middle two Quartile ranges and represents the 50% Confidence Interval between the 25% and 75% Confidence Levels for the data.

Interquintile Range   The Interquintile Range comprises the middle three Quintile ranges and represents the 60% Confidence Interval between the 20% and 80% Confidence Levels for the data.

Jarque-Bera Test   The Jarque-Bera Test is a statistical test for whether data can be assumed to follow a Normal Distribution. It exploits the properties of a Normal Distribution's Skewness and Excess Kurtosis being zero.

Kendall's Tau Rank Correlation Coefficient   Kendall's Tau Rank Correlation Coefficient for two variables is a statistic that measures the difference between the number of Concordant and Discordant data pairs as a proportion of the total number of possible unique pairings, where two pairs are said to be concordant if the ranks of the two variables move in the same direction, or are said to be discordant if the ranks of the two variables move in opposite directions.

Laspeyres Index   Laspeyres Indices are time-based indices which compare the prices of commodities at a point in time with the equivalent prices for the Index Base Period, based on the original quantities consumed at the Index Base Year.

Learning Curve   A Learning Curve is a mathematical representation of the degree at which the cost, time or effort to perform one or more activities reduces through the acquisition and application of knowledge and experience through repetition and practice.

Learning Curve Breakpoint   A Learning Curve Breakpoint is the position in the build or repetition sequence at which the empirical or theoretical rate of learning changes.

Learning Curve Cost Driver   A Learning Curve Cost Driver is an independent variable which affects or indicates the rate or amount of learning observed.

Learning Curve Segmentation   Learning Curve Segmentation refers to a technique which models the impact of discrete Learning Curve Cost Drivers as a product of multiple unit-based learning curves.

Learning Curve Step-point   A Learning Curve Step-point is the position in the build or repetition sequence at which there is a step function increase or decrease in the level of values evident on the empirical or theoretical Learning Curve.

Learning Exponent   A Learning Exponent is the power function exponent of a Learning Curve reduction and is calculated as the Logarithmic value of the Learning Rate using a Logarithmic Base equivalent to the Learning Rate Multiplier.

Learning Rate and Learning Rate Multiplier   The Learning Rate expresses the complement of the percentage reduction over a given Learning Rate Multiplier (usually 2). For example, an 80% Learning Rate with a Learning Multiplier of 2 implies a 20% reduction every time the quantity doubles.

Least Squares Regression   Least Squares Regression is a Regression procedure which identifies the 'Best Fit' of a pre-defined functional form by minimising the Sum of the Squares of the vertical difference between each data observation and the assumed functional form through the Arithmetic Mean of the data.

Leptokurtotic or Leptokurtic   An expression that the degree of Excess Kurtosis in a probability distribution is peakier than a Normal Distribution.

Linear Function   A Linear Function of two variables is one which can be represented as a monotonic increasing or decreasing Straight Line without any need for Mathematical Transformation.

Logarithm   The Logarithm of any positive value for a given positive Base Number not equal to one is that power to which the Base Number must be raised to get the value in question.

Logarithmic Function   A Logarithmic Function of two variables is one in which the dependent variable on the vertical axis produces a monotonic increasing or decreasing Straight Line, when plotted against the Logarithm of the independent variable on the horizontal axis.

Mann-Whitney U-Test sometimes known as Mann-Whitney-Wilcoxon U-Test   A U-Test is used to test whether two samples could be drawn from the same population by comparing the distribution of the joint ranks across the two samples.

Marching Army Technique sometimes referred to as Standing Army Technique   The Marching Army Technique refers to a technique that assumes that costs vary directly in proportion with a schedule.

Mathematical Transformation   A Mathematical Transformation is a numerical process in which the form, nature or appearance of a numerical expression is converted into an equivalent but non-identical numerical expression with a different form, nature or appearance.

Maximum   The Maximum is the largest observed value in a sample of data, or the largest potential value in a known or assumed statistical distribution. In some circumstances, the term may be used to imply a pessimistic value at the upper end of potential values rather than an absolute value.

Mean Absolute Deviation   See Average Absolute Deviation (AAD).

Measures of Central Tendency   Measures of Central Tendency is a collective term that refers to those descriptive statistics that measure key attributes of a data sample (Means, Modes and Median).

Measures of Dispersion and Shape   Measures of Dispersion and Shape is a collective term that refers to those descriptive statistics that measure the degree and/or pattern of scatter in the data in relation to the Measures of Central Tendency.

Median   The Median of a set of data is that value which occurs in the middle of the sequence when its values have been arranged in ascending or descending order. There are an equal number of data points less than and greater than the Median.

Median Absolute Deviation (MAD)   The Median Absolute Deviation of a range of data is the Median of the 'absolute' distance of each data point from the Median of those data points, ignoring the "sign" depicting whether each point is less than or greater than the Median.

Memoryless Probability Distribution   In relation to Queueing Theory, a Memoryless Probability Distribution is one in which the probability of waiting a set period of time is independent of how long we have been waiting already. The probability of waiting longer than the sum of two values is the product of the probabilities of waiting longer than each value in turn. An Exponential Distribution is the only Continuous Probability Distribution that exhibits this property, and a Geometric Distribution is the only discrete form.

Mesokurtotic or Mesokurtic   An expression that the degree of Excess Kurtosis in a probability distribution is comparable with a Normal Distribution.

Method   See Estimating Method.

Metric   A Metric is a statistic that measures an output of a process or a relationship between a variable and another variable or some reference point.
See also Estimating Metric.

Minimum   The Minimum is the smallest observed value in a sample of data, or the smallest potential value in a known or assumed statistical distribution. In some circumstances, the term may be used to imply an optimistic value at the lower end of potential values rather than an absolute value.

Mode   The Mode of a set of data is that value which has occurred most frequently, or that which has the greatest probability of occurring.

Model Validation and Verification   See individual terms: Validation and Verification.

Monotonic Function   A Monotonic Function of two paired variables is one that when values are arranged in ascending numerical order of one variable, the value of the other variable either perpetually increases or perpetually decreases.

Monte Carlo Simulation   Monte Carlo Simulation is a technique that models the range and relative probabilities of occurrence, of the potential outcomes of a number of input variables whose values are uncertain but can be defined as probability distributions.

Moving Average   A Moving Average is a series or sequence of successive averages calculated from a fixed number of consecutive input values that have occurred in a natural sequence. The fixed number of consecutive input terms used to calculate each average term is referred to as the Moving Average Interval or Base.

Moving Geometric Mean   A Moving Geometric Mean is a series or sequence of successive geometric means calculated from a fixed number of consecutive input values that have occurred in a natural sequence. The fixed number of consecutive input terms used to calculate each geometric mean term is referred to as the Moving Geometric Mean Interval or Base.

Moving Harmonic Mean   A Moving Harmonic Mean is a series or sequence of successive harmonic means calculated from a fixed number of consecutive input values that have occurred in a natural sequence. The fixed number of consecutive input terms used to calculate each harmonic mean term is referred to as the Moving Harmonic Mean Interval or Base.

Moving Maximum   A Moving Maximum is a series or sequence of successive maxima calculated from a fixed number of consecutive input values that have occurred in a natural sequence. The fixed number of consecutive input terms used to calculate each maximum term is referred to as the Moving Maximum Interval or Base.

Moving Median   A Moving Median is a series or sequence of successive medians calculated from a fixed number of consecutive input values that have occurred in a natural sequence. The fixed number of consecutive input terms used to calculate each median term is referred to as the Moving Median Interval or Base.

Moving Minimum   A Moving Minimum is a series or sequence of successive minima calculated from a fixed number of consecutive input values that have occurred in a natural sequence. The fixed number of consecutive input terms used to calculate each minimum term is referred to as the Moving Minimum Interval or Base.

Moving Standard Deviation   A Moving Standard Deviation is a series or sequence of successive standard deviations calculated from a fixed number of consecutive input values that have occurred in a natural sequence. The fixed number of consecutive input terms used to calculate each standard deviation term is referred to as the Moving Standard Deviation Interval or Base.

Multicollinearity   See Collinearity.

Multiplicative Time Series Model   See Time Series Analysis.

Multi-Variant Unit Learning   Multi-Variant Unit Learning is a technique that considers shared and unique learning across multiple variants of the same or similar recurring products.

Net Present Value The Net Present Value (NPV) of an investment is the sum of all positive and negative cash flows through time, each of which have been discounted based on the time value of money relative to a Base Year (usually the present year).

Nominal Year Values 'Nominal Year Values' are historical values expressed in terms of those that were current at the historical time at which they were incurred. In some cases, these may be referred to as 'Current Year Values'.

Norden-Rayleigh Curve A Norden-Rayleigh is an empirical relationship that models the distribution of resource required in the non-recurring concept demonstration or design and development phases.

Null Hypothesis A Null Hypothesis is that supposition that the difference between an observed value or effect and another observed or assumed value or effect, can be legitimately attributable to random sampling or experimental error. It is usually denoted as H_0.

Open Interval An Open Continuous Interval is one which excludes its endpoints, and is usually depicted with rounded brackets: (Minimum, Maximum).

Opportunity An Opportunity is an event or set of circumstances that may or may not occur, but if it does occur an Opportunity will have a beneficial effect on our plans, impacting positively on the cost, quality, schedule, scope compliance and/or reputation of our project or organisation.

Optimism Bias Optimism Bias is an expression of the inherent bias (often unintended) in an estimate output based on either incomplete or misunderstood input assumptions.

Outlier An outlier is a value that falls substantially outside the pattern of other data. The outlier may be representative of unintended atypical factors or may simply be a value which has a very low probability of occurrence.

Outturn Year Values 'Outturn Year Values' are values that have been adjusted to express an expectation of what might be incurred in the future due to escalation or other predicted changes. In some cases, these may be referred to as 'Then Year Values'.

Paasche Index Paasche Indices are time-based indices which compare prices of commodities at a point in time with the equivalent prices for the Index Base Period, based on the quantities consumed at the current point in time in question

Parametric Estimating Method A Parametric Estimating Method is a systematic means of establishing and exploiting a pattern of behaviour between the variable that we want to estimate, and some other independent variable or set of variables or characteristics that have an influence on its value.

Payback Period The Payback Period is an expression of how long it takes for an investment opportunity to break even, i.e. to pay back the investment.

Pearson's Linear Correlation Coefficient Pearson's Linear Correlation Coefficient for two variables is a measure of the extent to which a change in the value of one variable can be associated with a change in the value of the other variable through a linear relationship. As such it is a measure of linear dependence or linearity between the two variables, and can be calculated by dividing the Covariance of the two variables by the Standard Deviation of each variable.

Peirce's Criterion A test for multiple Outliers based on the deviation Z-Score of the suspect data point.

Percentile A Percentile is one of a hundred subsets from a set of ordered values which each nominally contain a hundredth of the total number of values in each subset. The term can also be used to express the values that divide the ordered values into the hundred ordered subsets.

Planned Value (PV) See Earned Value Management Abbreviations and Terminology.

Platykurtotic or Platykurtic   An expression that the degree of Excess Kurtosis in a probability distribution is shallower than a Normal Distribution.

Power Function   A Power Function of two variables is one in which the Logarithm of the dependent variable on the vertical axis produces a monotonic increasing or decreasing Straight Line when plotted against the Logarithm of the independent variable on the horizontal axis.

Precision
  (1)   Precision is an expression of how close repeated trials or measurements are to each other.
  (2)   Precision is an expression of the level of exactness reported in a measurement, statistic or estimate.

Primary Data   See Data Type.

Primary Driver   See Estimate Drivers.

Probability Density Function (PDF)   The Probability Density Function of a Continuous Random Variable expresses the rate of change in the probability distribution over the range of potential continuous values defined, and expresses the relative likelihood of getting one value in comparison with another.

Probability Mass Function (PMF)   The Probability Mass Function of a Discrete Random Variable expresses the probability of the variable being equal to each specific value in the range of all potential discrete values defined. The sum of these probabilities over all possible values equals 100%.

Probability of Occurrence   A Probability of Occurrence is a quantification of the likelihood that an associated Risk or Opportunity will occur with its consequential effects.

Quadratic Mean or Root Mean Square   The Quadratic Mean of a set of n numerical data values is a statistic calculated by taking the square root of the Arithmetic Mean of the squares of the n values. As a consequence, it is often referred to as the Root Mean Square.

Quantile   A Quantile is the generic term for a number of specific measures that divide a set of ordered values into a quantity of ranges with an equal proportion of the total number of values in each range. The term can also be used to express the values that divide the ordered values into such ranges.

Quantity-based Learning Curve   A Quantity-based Learning Curve is an empirical relationship which reflects that the time, effort or cost to perform an activity reduces as the number of repetitions of that activity increases.

Quartile   A Quartile is one of four subsets from a set of ordered values which nominally contain a quarter of the total number of values in each subset. The term can also be used to express the values that divide the ordered values into the four ordered subsets.

Queueing Theory   Queueing Theory is that branch of Operation Research that studies the formation and management of queuing systems and waiting times.

Quintile   A Quintile is one of five subsets from a set of ordered values which nominally contain a fifth of the total number of values in each subset. The term can also be used to express the values that divide the ordered values into the five ordered subsets.

Range   The Range is the difference between the Maximum and Minimum observed values in a dataset, or the Maximum and Minimum theoretical values in a statistical distribution. In some circumstances, the term may be used to imply the difference between pessimistic and optimistic values from the range of potential values rather than an absolute range value.

Rate Metric   A Rate is an Estimating Metric used to quantify how one variable's value changes in relation to some measurable driver, attribute or parameter, and would be expressed in the form of a [Value] of one attribute per [Unit] of another attribute.

Ratio Metric   A Ratio is an Estimating Metric used to quantify the relative size proportions between two different instances of the same driver, attribute or characteristic such as weight. It is typically used as an element of Estimating by Analogy or in the Normalisation of data.

Real Year Values   'Real Year Values' are values that have been adjusted to take account of historical or future inflationary effects or other changes, and are expressed in relation to the Current Year Values for any defined year. They are often referred to as 'Constant Year Values'.

Regression Analysis   Regression Analysis is a systematic procedure for establishing the Best Fit relationship of a predefined form between two or more variables, according to a set of Best Fit criteria.

Regression Confidence Interval   The Regression Confidence Interval of a given probability is an expression of the Uncertainty Range around the Regression Line. For a known value of a single independent variable, or a known combination of values from multiple independent variables, the mean of all future values of the dependent variable will occur within the Confidence Interval with the probability specified.

Regression Prediction Interval   A Regression Prediction Interval of a given probability is an expression of the Uncertainty Range around future values of the dependent variable based on the regression data available. For a known value of a single independent variable, or a known combination of values from multiple independent variables, the future value of the dependent variable will occur within the Prediction Interval with the probability specified.

Residual Risk Exposure   The Residual Risk Exposure is the weighted value of the Risk, calculated by multiplying its Most Likely Value by the complement of its Probability of Occurrence (100% − Probability of Occurrence). It is used to highlight the relative value of the risk that is not covered by Risk Exposure calculation.

Risk   A Risk is an event or set of circumstances that may or may not occur, but if it does occur a Risk will have a detrimental effect on our plans, impacting negatively on the cost, quality, schedule, scope compliance and/or reputation of our project or organisation.

Risk Exposure   A Risk Exposure is the weighted value of the Risk, calculated by multiplying its Most Likely Value by its Probability of Occurrence.
See also Residual Risk Exposure.

Risk & Opportunity Ranking Factor   A Risk & Opportunity Ranking Factor is the relative absolute exposure of a Risk or Opportunity in relation to all others, calculated by dividing the absolute value of the Risk Exposure by the sum of the absolute values of all such Risk Exposures.

Risk Uplift Factors   A Top-down Approach to Risk Analysis may utilise Risk Uplift Factors to quantify the potential level of risk based on either known risk exposure for the type of work being undertaken based on historical records of similar projects, or based on a Subject Matter Expert's Judgement.

R-Square (Regression)   R-Square is a measure of the "Goodness of Fit" of a simple linear regression model to a set of data points. It is directly equivalent to the Coefficient of Determination that shows how much of the total variance in one variable can be explained by the variance in the other variable.
See also Adjusted R-Square.

Schedule Maturity Assessment (SMA)   A Schedule Maturity Assessment provides a 'health warning' on the maturity of a schedule based on its underpinning assumptions and interdependencies, and takes account of the level of task definition available and historical evidence used.

Secondary Data   See Data Type.

Secondary Driver   See Estimate Drivers.

Skewness Coefficient   The Fisher-Pearson Skewness Coefficient is an expression of the degree of asymmetry of a set of values around their Arithmetic Mean. A positive Skewness Coefficient indicates that the data has a longer tail on the right-hand side, in the direction of the positive axis; such data is said to be Right or Positively Skewed. A negative Skewness Coefficient indicates that the data has a longer tail on the left-hand side, in the direction of the negative axis; such data is said to be Left or Negatively Skewed. Data that is distributed symmetrically returns a Skewness Coefficient of zero.

Slipping and Sliding Technique   A technique that compares and contrasts a Bottom-up Monte Carlo Simulation Cost evaluation of Risk, Opportunity and Uncertainty with a holistic Top-down Approach based on Schedule Risk Analysis and Uplift Factors.

Spearman's Rank Correlation Coefficient   Spearman's Rank Correlation Coefficient for two variables is a measure of monotonicity of the ranks of the two variables, i.e. the degree to which the ranks move in the same or opposite directions consistently. As such it is a measure of linear or non-linear interdependence.

SPI (Schedule Performance Index – Cost Impact)   See Earned Value Management Abbreviations and Terminology.

SPI(t) (Schedule Performance Index – Time Impact)   See Earned Value Management Abbreviations and Terminology.

Spreadsheet Validation and Verification   See individual terms: Validation and Verification.

Standard Deviation of a Population   The Standard Deviation of an entire set (population) of data values is a measure of the extent to which the data is dispersed around its Arithmetic Mean. It is calculated as the square root of the Variance, which is the average of the squares of the deviations of each individual value from the Arithmetic Mean of all the values.

Standard Deviation of a Sample   The Standard Deviation of a sample of data taken from the entire population is a measure of the extent to which the sample data is dispersed around its Arithmetic Mean. It is calculated as the square root of the Sample Variance, which is the sum of squares of the deviations of each individual value from the Arithmetic Mean of all the values divided by the degrees of freedom, which is one less than the number of data points in the sample.

Standard Error   The Standard Error of a sample's statistic is the Standard Deviation of the sample values of that statistic around the true population value of that statistic. It can be approximated by the dividing the Sample Standard Deviation by the square root of the sample size.

Stanford-B Unit Learning Curve   A Stanford-B Unit Learning Curve is a variation of the Crawford Unit Learning Curve that allows for the benefits of prior learning to be expressed in terms of an adjustment to the effective number of cumulative units produced.

Statistics

(1)   The science or practice relating to the collection and interpretation of numerical and categorical data for the purposes of describing or inferring representative values of the whole data population from incomplete samples.

(2) The numerical values, measures and context that have been generated as outputs from the above practice.

Stepwise Regression Stepwise Regression by Forward Selection is a procedure by which a Multi-Linear Regression is compiled from a list of independent candidate variables, commencing with the most statistically significant individual variable (from a Simple Linear Regression perspective) and progressively adding the next most significant independent variable, until such time that the addition of further candidate variables does not improve the fit of the model to the data in accordance with the accepted Measures of Goodness of Fit for the Regression.

Stepwise Regression by Backward Elimination is a procedure by which a Multi-Linear Regression is compiled commencing with all potential independent candidate variables and eliminating the least statistically significant variable progressively (one at a time) until such time that all remaining candidate variables are deemed to be statistically significant in accordance with the accepted Measures of Goodness of Fit.

Subject Matter Expert's Opinion (Expert Judgement) Expert Judgement is a recognised term expressing the opinion of a Subject Matter Expert (SME).

SV (Schedule Variance – Cost Impact) See Earned Value Management Abbreviations and Terminology.

SV(t) (Schedule Variance – Time Impact) See Earned Value Management Abbreviations and Terminology.

Tertiary Data See Data Type.

Then Year Values 'Then Year Values' are values that have been adjusted to express an expectation of what might be incurred in the future due to escalation or other predicted changes. In some cases, these may be referred to as 'Outturn Year Values'.

Three-Point Estimate See 3-Point Estimate.

Time Series Analysis Time Series Analysis is the procedure whereby a series of values obtained at successive time intervals is separated into its constituent elements that describe and calibrate a repeating pattern of behaviour over time in relation to an underlying trend.

An Additive/Subtractive Time Series Model is one in which the Predicted Value is a function of the forecast value attributable to the underlying Trend plus or minus adjustments for its relative Seasonal and Cyclical positions in time.

A Multiplicative Time Series Model is one in which the Predicted Value is a function of the forecast value attributable to the underlying Trend multiplied by appropriate Seasonal and Cyclical Factors.

Time-Based Learning Curve A Time-based Learning Curve is an empirical relationship which reflects that the time, effort or cost to produce an output from an activity decreases as the elapsed time since commencement of that activity increases.

Time-Constant Learning Curve A Time-Constant Learning Curve considers the output or yield per time period from an activity rather than the time or cost to produce a unit. The model assumes that the output increases due to learning, from an initial starting level, before flattening out asymptotically to a steady state level.

Time-Performance Learning Curve A Time-Performance Learning Curve is an empirical relationship that expresses the reduction in the average time or cost per unit produced per period as a power function of the cumulative number periods since production commenced.

Top-down Approach   In a top-down approach to estimating, the estimator reviews the overall scope of work in order to identify the major elements of work and characteristics (drivers) that could be estimated separately from other elements. Typically, the estimator might consider a natural flow down through the Work Breakdown Structure (WBS), Product Breakdown Structure (PBS) or Service Breakdown Structure (SBS). The estimate scope may be broken down to different levels of WBS etc as required; it is not necessary to cover all elements of the task at the same level, but the overall project scope must be covered. The overall project estimate would be created by aggregating these high-level estimates. Lower level estimates can be created by subsequent iterations of the estimating process when more definition becomes available, and bridging back to the original estimate.

TRACEability   A Basis of Estimate should satisfy the principles of TRACEability:

Transparent – clear and unambiguous with nothing hidden

Repeatable – allowing another estimator to reproduce the same results with the same information

Appropriate – it is justifiable and relevant in the context it is to be used

Credible – it is based on reality or a pragmatic reasoned argument that can be understood and is believable

Experientially-based – it can be underpinned by reference to recorded data (evidence) or prior confirmed experience

Transformation   See Mathematical Transformation.

Trusted Source Estimating Method   The Trusted Source Method of Estimating is one in which the Estimate Value is provided by a reputable, reliable or undisputed source. Typically, this might be used for low value cost elements. Where the cost element is for a more significant cost value, it would not be unreasonable to request the supporting Basis of Estimate, but this may not be forthcoming if the supporting technical information is considered to be proprietary in nature.

t-Test   A t-Test is used for small sample sizes (< 30) to test probability of getting a sample's test statistic (often the Mean), if the equivalent population statistic has an assumed different value. It is also used to test whether two samples could be drawn from the same population.

Tukey's Fences   A test for a single Outlier based on the Inter-Quartile Range of the data sample.

Type I Error   A Type I Error is one in which we accept a hypothesis we should have rejected.

Type II Error   A Type II Error is one in which we reject a hypothesis we should have accepted.

U-Test   See Mann-Whitney U-Test.

Uncertainty   Uncertainty is an expression of the lack of exactness around a variable, and is frequently quantified in terms of a range of potential values with an optimistic or lower end bound and a pessimistic or upper end bound.

Validation (Spreadsheet or Model)   Validation is the process by which the assumptions and data used in a spreadsheet or model are checked for accuracy and appropriateness for their intended purpose.

See also Verification.

Variance of a Population   The Variance of an entire set (population) of data values is a measure of the extent to which the data is dispersed around its Arithmetic Mean. It is calculated as the average of the squares of the deviations of each individual value from the Arithmetic Mean of all the values.

Variance of a Sample The Variance of a Sample of data taken from the entire population is a measure of the extent to which the sample data is dispersed around its Arithmetic Mean. It is calculated as the sum of squares of the deviations of each individual value from the Arithmetic Mean of all the values divided by the degrees of freedom, which is one less than the number of data points in the sample.

Verification (Spreadsheet or Model) Verification is the process by which the calculations and logic of a spreadsheet or model are checked for accuracy and appropriateness for their intended purpose.

See also Validation.

Wilcoxon-Mann-Whitney U-Test See Mann-Whitney U-Test.

Wright's Cumulative Average Learning Curve Wright's Cumulative Average Learning Curve is an empirical relationship that expresses the reduction in the cumulative average time or cost of each unit produced as a power function of the cumulative number units produced.

Z-Score A Z-Score is a statistic which standardises the measurement of the distance of a data point from the Population Mean by dividing by the Population Standard Deviation.

Z-Test A Z-Test is used for large sample sizes (< 30) to test probability of getting a sample's test statistic (often the Mean), if the equivalent population statistic has an assumed different value.

Legend for Microsoft Excel Worked Example Tables in Greyscale

Cell type	Potential Good Practice Spreadsheet Modelling Colour	Greyscale used in Book	Example of Greyscale Used in Book
Header or Label	Light Grey	Text on grey	Text
Constant	Deep blue	Bold white numeric on black	1
Input	Pale Yellow	Normal black numeric on pale grey	23
Calculation	Pale Green	Normal black numeric on mid grey	45
Solver variable	Lavender	Bold white numeric on mid grey	67
Array formula	Bright Green	Bold white numeric on dark grey	89
Random Number	Pink	Bold black numeric on dark grey	0.0902
Comment	White	Text on white	Text

Index

"In the Working Guides to Estimating and Forecasting Alan has managed to capture the full spectrum of relevant topics with simple explanations, practical examples and academic rigor, while injecting humour into the narrative."

– *Dale Shermon*, Chairman, Society of Cost Analysis and Forecasting (SCAF)

"If estimating has always baffled you, this innovative well illustrated and user friendly book will prove a revelation to its mysteries. To confidently forecast, minimise risk and reduce uncertainty we need full disclosure into the science and art of estimating. Thankfully, and at long last the "Working Guides to Estimating & Forecasting" are exactly that, full of practical examples giving clarity, understanding and validity to the techniques. These are comprehensive step by step guides in understanding the principles of estimating using experientially based models to analyse the most appropriate, repeatable, transparent and credible outcomes. Each of the five volumes affords a valuable tool for both corporate reference and an outstanding practical resource for the teaching and training of this elusive and complex subject. I wish I had access to such a thorough reference when I started in this discipline over 15 years ago, I am looking forward to adding this to my library and using it with my team."

– *Tracey L Clavell*, Head of Estimating & Pricing, BAE Systems Australia

"At last, a comprehensive compendium on these engineering math subjects, essential to both the new and established "cost engineer"! As expected the subjects are presented with the author's usual wit and humour on complex and daunting "mathematically challenging" subjects. As a professional trainer within the MOD Cost Engineering community trying to embed this into my students, I will be recommending this series of books as essential bedtime reading."

– *Steve Baker*, Senior Cost Engineer, DE&S MOD

"Alan has been a highly regarded member of the Cost Estimating and forecasting profession for several years. He is well known for an ability to reduce difficult topics and cost estimating methods down to something that is easily digested. As a master of this communication he would most often be found providing training across the cost estimating and forecasting tools and at all levels of expertise. With this 5-volume set, *Working Guides to Estimating and Forecasting*, Alan has brought his normal verbal training method into a written form. Within their covers Alan steers away from the usual dry academic script into establishing an almost 1:1 relationship with the reader. For my money a recommendable read for all levels of the Cost Estimating and forecasting profession and those who simply want to understand what is in the 'blackbox' just a bit more."

– *Prof Robert Mills*, Margin Engineering, Birmingham City University.
MACOSTE, SCAF, ICEAA

"Finally, a book to fill the gap in cost estimating and forecasting! Although other publications exist in this field, they tend to be light on detail whilst also failing to cover many of the essential aspects of estimating and forecasting. Jones covers all this and more from both a theoretical and practical point of view, regularly drawing on his considerable experience in the defence industry to provide many practical examples to support his comments. Heavily illustrated throughout, and often presented in a humorous fashion, this is a must read for those who want to understand the importance of cost estimating within the broader field of project management."

– *Dr Paul Blackwell*, Lecturer in Management of Projects, The University of Manchester, UK

"Alan Jones provides a useful guidebook and navigation aid for those entering the field of estimating as well as an overview for more experienced practitioners. His humorous asides supplement a thorough explanation of techniques to liven up and illuminate an area which has little attention in the literature, yet is the basis of robust project planning and successful delivery. Alan's talent for explaining the complicated science and art of estimating in practical terms is testament to his knowledge of the subject and to his experience in teaching and training."

– *Therese Lawlor-Wright*, Principal Lecturer in Project Management at the University of Cumbria

"Alan Jones has created an in depth guide to estimating and forecasting that I have not seen historically. Anyone wishing to improve their awareness in this field should read this and learn from the best."

– *Richard Robinson*, Technical Principal for Estimating, Mott MacDonald

"The book series of 'Working Guides to Estimating and Forecasting' is an essential read for students, academics and practitioners who interested in developing a good understanding of cost estimating and forecasting from real-life perspectives."

– *Professor Essam Shehab*, Professor of Digital Manufacturing and Head of Cost Engineering, Cranfield University, UK

"In creating the *Working Guides to Estimating and Forecasting,* Alan has captured the core approaches and techniques required to deliver robust and reliable estimates in a single series. Some of the concepts can be challenging, however, Alan has delivered them to the reader in a very accessible way that supports lifelong learning. Whether you are an apprentice, academic or a seasoned professional, these working guides will enhance your ability to understand the alternative approaches to generating a well-executed, defensible estimate, increasing your ability to support competitive advantage in your organisation."

– *Professor Andrew Langridge*, Royal Academy of Engineering Visiting Professor in Whole Life Cost Engineering and Cost Data Management, University of Bath, UK

"Alan Jones's "*Working Guides to Estimating and Forecasting*" provides an excellent guide for all levels of cost estimators from the new to the highly experienced. Not only does he cover the underpinning good practice for the field, his books will take you on a journey from cost estimating basics through to how estimating should be used in manufacturing the future – reflecting on a whole life cycle approach. He has written a must-read book for anyone starting cost estimating as well as for those who have been doing estimates for years. Read this book and learn from one of the best."

– *Linda Newnes*, Professor of Cost Engineering, University of Bath, UK